T0259009

Innovation in PMBOK through Industrial Revolution 4.0

In this textbook for upper-undergraduate and postgraduate students, Dr Ali and colleagues provide the reader with information on the effect of Industrial Revolution 4.0 on the construction industry, particularly regarding PMBOK knowledge areas.

The authors furnish readers with an understanding of IR 4.0 and reasons for adopting it and provide an in-depth insight into the impact of IR 4.0 on technology and society, particularly in the construction industry. Further to this, they also compare traditional and IR 4.0 project manager's competencies so that readers can develop their understanding of Project Management Knowledge areas and how IR 4.0 can be used to enhance these knowledge areas. The book is structured logically and sequentially to benefit even novice readers as they progress from basic to more advanced topics related to IR 4.0 and PMBOK. The final main chapter of this book provides an in-depth discussion of the enhancement of PMBOK knowledge areas using IR 4.0, including topics such as project integration management, IR 4.0 enhancements such as digitalization, and a conceptual framework for industry betterment. By the end of the book, readers will have the knowledge and skills to apply IR 4.0 techniques in their future careers in the construction industry.

This book is an invaluable resource for students of construction engineering and management at upper-undergraduate and post-graduate levels and for existing industry professionals as part of their continuous professional development.

Muhammad Ali Musarat is a civil engineer, with a specialization in Construction Engineering and Management. He is currently pursuing his career as a postdoctoral researcher at the Department of Civil and Environmental Engineering, Universiti Teknologi PETRONAS, Malaysia.

Muhammad Irfan is a lecturer at the Department of Civil Engineering, HITEC University, Pakistan. He is currently pursuing his PhD degree from COMSATS University Islamabad, Pakistan. He obtained his bachelor's and master's degrees in 2016 and 2019, respectively.

Maria Ghufran is a researcher with the Department of Construction Engineering and Management, National University of Sciences and Technology, Islamabad, Pakistan.

Wesam Salah Alaloul is a senior lecturer at Universiti Teknologi PETRONAS. He is also a certified trainer for construction costing and sustainability issues. He obtained his bachelor's degree, master's degree, and PhD in 2010, 2012, and 2017, respectively.

Innovation in PMBOK through Industrial Revolution 4.0

An Automated Solution for the Construction Industry

Muhammad Ali Musarat, Muhammad Irfan, Maria Ghufran, and Wesam Salah Alaloul

CRC Press
Taylor & Francis Group
Boca Raton London New York

CRC Press is an imprint of the
Taylor & Francis Group, an **informa** business

Designed cover image: Industry 4.0 technology concept - smart factory for the fourth industrial revolution

First edition published 2024
by CRC Press
2385 NW Executive Center Drive, Suite 320, Boca Raton FL 33431

and by CRC Press
4 Park Square, Milton Park, Abingdon, Oxon, OX14 4RN

CRC Press is an imprint of Taylor & Francis Group, LLC

ISBN: 978-1-032-62174-6 (hbk)
ISBN: 978-1-032-62175-3 (pbk)
ISBN: 978-1-032-62176-0 (ebk)

DOI: 10.1201/9781032621760

Typeset in Sabon
by Apex CoVantage, LLC

Contents

CHAPTER 1 ▪ Introduction   1

1.1 THREE-TIERED IMPACT OF TECHNOLOGICAL
     ADVANCEMENT   1

     1.1.1 Society   2

     1.1.2 Economic   3

     1.1.3 Environment   4

1.2 THE INTERRELATIONSHIP BETWEEN TECHNOLOGY
     AND THE INDUSTRIAL REVOLUTION   4

1.3 PROJECT MANAGEMENT IN THE ERA OF THE
     INDUSTRIAL REVOLUTION   5

CHAPTER 2 ▪ Concept of Industry 4.0   8

2.1 INDUSTRY 4.0 OR IR 4.0   8

2.2 REASONS TO ADOPT INDUSTRY 4.0   11

2.3 STRATEGIES IN IR 4.0   19

     2.3.1 In-Use Technologies of IR 4.0   20

     2.3.2 IR 4.0 Technologies That Are Not Fully
            Available for Use   21

2.4 INDUSTRY 4.0 IMPACTS ON TECHNOLOGY   24

2.5 INDUSTRY 4.0 IMPACTS ON SOCIETY   26

     2.5.1 Negative Impacts   26

     2.5.2 Positive Impacts   27

2.6 INDUSTRY 4.0 IMPACTS ON THE ENVIRONMENT   30

2.7   INDUSTRY 4.0 IMPACTS ON THE WORKPLACE   32

REFERENCES   33

CHAPTER 3 ■ Industrial Revolution and Construction
                  Industry   37

3.1   MANUFACTURING AND CONSTRUCTION
        INDUSTRIES IN THE CONTEXT OF IR 4.0   37

3.2   IMPLEMENTATION OF INDUSTRY REVOLUTION
        IN THE CONSTRUCTION INDUSTRY   39

        3.2.1   Emerging IR 4.0 Technologies for the
                  Construction Industry   39

        3.2.2   Incorporation of IR 4.0 in Construction   59

        3.2.3   Intelligent Construction Hurdles   60

3.3   CONSTRUCTION 4.0 IMPLEMENTATION DRIVERS   62

        3.3.1   Construction 4.0 and Project Life Cycle   62

        3.3.2   Mindset Transition and Training   62

        3.3.3   Strategy Development by Organizations   63

        3.3.4   International Consensus on Construction 4.0   64

        3.3.5   Digitized Designs   65

3.4   CHALLENGES OF APPLICATION OF INDUSTRIAL
        REVOLUTION IN THE CONSTRUCTION INDUSTRY   65

        3.4.1   Problem of Cybersecurity   66

        3.4.2   Adoption and Maintenance of IR 4.0
                  Technologies   67

        3.4.3   Hesitation of Organizations and Inadequate
                  Supportive Facilities   68

        3.4.4   Shortage of Skilled Workforce   68

        3.4.5   Insufficient Technical Competence and the
                  Unavailability of Laws and Regulations   68

        3.4.6   Accountability, Transparency, and
                  Privacy Issues   68

        3.4.7   Lack of Interoperability and Standards   69

REFERENCES   70

CHAPTER 4 ■ Project Management Knowledge Areas   71

4.1   PROJECT MANAGEMENT PHILOSOPHY
IMPLEMENTATION IN CONSTRUCTION   71

4.2   IR 4.0 AS A SOLUTION FOR PROJECT MANAGEMENT
KNOWLEDGE AREAS   73

    4.2.1   Traditional Project Management Skills   73

    4.2.2   Industry 4.0 Project Management Skills   75

    4.2.3   Traditional vs. Industry 4.0 Project Manager's
Competencies   77

    4.2.4   Intelligent and Smart Manufacturing   82

    4.2.5   Intelligent and Smart Supply Chain
Management   88

REFERENCES   88

CHAPTER 5 ■ IR 4.0 Integration with PMBOK and
Sustainability   89

5.1   CONCEPT OF SUSTAINABILITY   89

5.2   SUSTAINABILITY AND PROJECT MANAGEMENT   96

5.3   IR 4.0 AND ECONOMIC SUSTAINABILITY 4.0   99

5.4   IR 4.0 AND ENVIRONMENTAL SUSTAINABILITY 4.0   102

5.5   IR 4.0 AND SOCIAL SUSTAINABILITY 4.0   105

5.6   ENHANCEMENT   109

5.7   DISCUSSION   112

REFERENCES   114

CHAPTER 6 ■ Project Management and Sustainable
Buildings in the Era of IR 4.0   119

6.1   SUSTAINABLE BUILDINGS   119

6.2   ROLE OF SUSTAINABLE BUILDINGS IN
ACHIEVING SUSTAINABILITY   120

    6.2.1   Enhanced Energy Efficiency   122

    6.2.2   GHG Reductions   123

    6.2.3   Improved Environment for Occupants   127

6.2.4	Resource Efficiency	130
6.2.5	Design and Management Efficiency	132
6.2.6	Effective Land Use	133
6.2.7	Water Conservation Efficiency	136
6.3	SUSTAINABLE BUILDING MANAGEMENT	138
6.4	SUSTAINABLE BUILDINGS IN THE ERA OF IR 4.0 FOR ENHANCING THE PROJECT MANAGEMENT	143
6.5	CASE STUDIES	146
6.5.1	Bahrain World Trade Center, Bahrain	147
6.5.2	Museum of Tomorrow, Brazil	147
6.5.3	Pixel Building, Australia	147
6.5.4	Bullitt Center, United States	147
6.5.5	The Crystal, United Kingdom	147
6.5.6	Sun-Moon Mansion, China	148
6.5.7	Suzlon One Earth, Pune, India	148
6.5.8	One Angel Square, Manchester, UK	148
6.5.9	The Edge, Amsterdam, Netherlands	148
6.5.10	One Central Park, Australia	149
REFERENCE		149

CHAPTER 7 ◼ Enhancement to PMBOK Knowledge Areas with IR 4.0 Perspective — 150

7.1	URBANIZATION AND I.R 4.0	150
7.2	PROJECT INTEGRATION MANAGEMENT	152
7.2.1	Conventional Way of Dealing	153
7.2.2	Advancements through IR 4.0	153
7.3	PROJECT SCOPE MANAGEMENT	155
7.3.1	Conventional Way of Dealing	155
7.3.2	Advancements through IR 4.0	156
7.4	PROJECT TIME MANAGEMENT	156
7.4.1	Conventional Way of Dealing	158
7.4.2	Advancements through IR 4.0	158

7.5   PROJECT COST MANAGEMENT   160

       7.5.1   Conventional Way of Dealing   161

       7.5.2   Advancements through IR 4.0   161

7.6   PROJECT QUALITY MANAGEMENT   162

       7.6.1   Conventional Way of Dealing   163

       7.6.2   Advancements through IR 4.0   163

7.7   PROJECT HUMAN RESOURCE MANAGEMENT   165

       7.7.1   Conventional Way of Dealing   166

       7.7.2   Advancements through IR 4.0   166

7.8   PROJECT COMMUNICATION MANAGEMENT   167

       7.8.1   Conventional Way of Dealing   168

       7.8.2   Advancements through IR 4.0   168

7.9   PROJECT PROCUREMENT MANAGEMENT   170

       7.9.1   Conventional Way of Dealing   170

       7.9.2   Advancements through IR 4.0   171

7.10   PROJECT RISK MANAGEMENT   171

       7.10.1 Conventional Way of Dealing   173

       7.10.2 Advancements through IR 4.0   173

7.11   PROJECT STAKEHOLDER MANAGEMENT   174

       7.11.1 Conventional Way of Dealing   175

       7.11.2 Advancements through IR 4.0   176

7.12   PROJECT SAFETY MANAGEMENT   177

       7.12.1 Conventional Way of Dealing   178

       7.12.2 Advancements through IR 4.0   178

7.13   PROJECT ENVIRONMENTAL MANAGEMENT   180

       7.13.1 Conventional Way of Dealing   181

       7.13.2 Advancements through IR 4.0   181

7.14   PROJECT FINANCIAL MANAGEMENT   182

       7.14.1 Conventional Way of Dealing   183

       7.14.2 Advancements through IR 4.0   184

7.15   ENHANCEMENT THROUGH IR 4.0   186

       7.15.1 Cyber-Physical System (CPS)   186

7.15.2 Cloud Systems (CS)   188

7.15.3 Machine-to-Machine (M2M)
Communication   189

7.15.4 Internet of Things (IoT) and Internet of
Services (IoS)   191

7.15.5 Big Data and Data Mining   193

7.15.6 Intelligent Robotics   195

7.15.7 Augmented Reality and Simulation   197

7.15.8 Enterprise Resource Planning (ERP) and
Business Intelligence   199

7.15.9 Smart Virtual Product Development
System (SVPD)   202

7.16 DISCUSSION   204

7.17 SUMMARY   208

REFERENCES   209

INDEX   215

CHAPTER 1

Introduction

THE ADVANCEMENT OF BOTH science and technology is fueled by one another. Building new technologies with scientific knowledge enables us to make novel findings about the world, which in turn enables us to expand our scientific knowledge base, which in turn motivates the development of new technologies, and so on. We believe that humans will continue to create novel inventions because we enjoy exploring. Fundamentally, technological development is our effort to better understand how things function, intending to create new procedures, goods, or services that will benefit society and make life easier. We utilize our cognitive, affective, and psychomotor capabilities to learn how things function and then apply that knowledge to create innovative technologies. Our environment is constantly changing, and if we don't learn to adapt, we will fall behind. Because of this, we continue to innovate and improve to remain competitive. This is an area where technology innovation is crucial. It enables us to increase our productivity so that we can simply complete daily duties and generate more revenue. We might utilize it to find answers to our current challenges and get rid of possible potential hazards. It offers us fresh approaches to carry out our daily tasks for improved outcomes.

1.1 THREE-TIERED IMPACT OF TECHNOLOGICAL ADVANCEMENT

The advancement in technology has a three-tier influence, as shown in Figure 1.1, that comprises environmental, societal, and economic impacts for the attainment of sustainability.

1.1.1 Society

Technology has an impact on how people interact, acquire knowledge, and think. It benefits society and impacts how people relate to one another regularly. Today's civilization is significantly influenced by technology. It affects people's daily lives and has both positive and harmful consequences on the planet. Modern times are

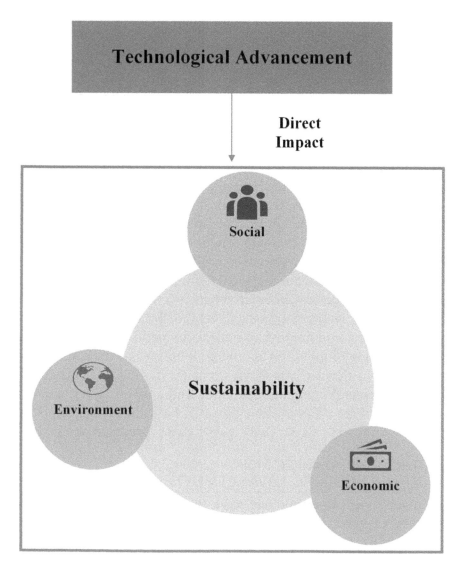

FIGURE 1.1   Impact of Technological Advancement.

characterized by frequent technical advancements. The way technological advances influence learning is one area where it has had a significant impact on society. Learning has become increasingly collaborative and interactive, which has improved how well people interact with the content they are learning and finding difficult. Additionally, it gives you improved accessibility to resources. Since the invention of the internet, we have had constant access to knowledge, and you can find nearly anything online. Communication, or how we speak and interact with one another globally, is another area where technology has had an impact on society. Thus, due to the abundance of quick and free instructional content made possible by technology, learning is now more individualized and self-paced.

1.1.2 Economic

By converting conventional procedures into digital workflows, the digitalization of industries—driven by technological advancements— allows companies to streamline operations, improve interactions with clients, and create novel business models. While traditional industries like media, retail, manufacturing, construction, and transportation have been disrupted by this change, it is also giving businesses new opportunities to use digital platforms, cloud computing, and data analytics to increase productivity, personalize offerings, and investigate novel revenue streams. Technology developments have accelerated the development of the digital economy, in which most transactions take place online. This has created new opportunities for economic activity and made it simpler for companies to contact clients and markets throughout the world. The single most important characteristic that separates advanced economies from prehistoric ones is the digital economy, which is also offering new chances for entrepreneurs, small enterprises, and established corporations to increase their reach and scale. Technology boosts both economic and distributive productivity for all businesses since it lowers production costs and offers new products. Technology development is uncertain, though, and nobody can predict what will be discovered or when or where. Simply gauging the pace of new technologies is exceedingly challenging. One should prepare and equip with the latest trends, tools, and methods by investing time and money towards learning new traits to move with the pace of the world.

1.1.3 Environment

Consequently, technology's impact on the environment can be either positive or detrimental, just like how it affects culture, society, and the economy. In other words, technology is a two-edged sword that has the power to both cause and repair damage to the environment. We see technological impacts on the environment under four key areas such as clean energy, water quality, waste reduction, and climate conservation. Therefore, the utilization of energy, the driving force behind industrialization, while preserving environmental quality remains a major concern for all the world's economies. However, technology has enabled substantial reductions in vehicle emissions of volatile organic compounds and carbon monoxide per mile traveled, driven by economic, legal, and environmental concerns. When energy use is brought up, the issue of possible climate change naturally arises that can be managed through the adoption of modern tools and techniques. Moreover, the two main environmental problems associated with energy usage and production are air quality and climate change, but they are not the only ones. Water quality is also harmed by industrial and vehicle pollutants, particularly nitrogen oxides. In water bodies such as rivers, lakes, and estuaries, nitrogen deposition serves as a fertilizer and encourages the growth of algae, resulting in eutrophic circumstances that damage submerged aquatic plants.

1.2 THE INTERRELATIONSHIP BETWEEN TECHNOLOGY AND THE INDUSTRIAL REVOLUTION

The Industrial Revolution (IR), a period of profound change, was greatly influenced by technological advancements. The journey of IR began from IR 1.0, followed by IR 2.0, IR 3.0, and IR 4.0. Thus, through innovations like the steam engine, power loom, and spinning jenny, traditionally labor-intensive operations were mechanized, which not only increased production but also completely altered how products were made. These advances in technology made it possible to produce goods in large quantities while also sharply cutting costs. The shipment of commodities over large distances was substantially facilitated by new transportation technology like the steam engine and improved roads, opening new markets and enlarging trade networks. Additionally, information could be shared between people and organizations more quickly because of communication technology like telegraphs, which promoted innovation and cooperation on a worldwide scale. In essence, the Industrial Revolution was sparked by the convergence of these technological advances, which forever altered how society views productivity, trade, and social structures.

The technological revolution and the feeling of constant change started far earlier than the eighteenth century and have lasted right up to the present. The IR's fusion of technology and industry was perhaps its most distinctive feature. Practically every area of human activity that exists now has been shaped along industrial lines by significant inventions and developments, and numerous new industries have also been created. However, technology is simply one component of IR 4.0, and companies must guarantee that their employees are adequately trained by upskilling and reskilling them, and then acquire new personnel as needed if they want to succeed in the fourth industrial revolution. Employees who are upskilled acquire new abilities to support them in their present roles, as the skills they require change.

1.3 PROJECT MANAGEMENT IN THE ERA OF THE INDUSTRIAL REVOLUTION

The exponential developments that have taken place since the emergence of global interconnectivity are impossible to overlook. The synergistic connection between humans and technology is causing unprecedented degrees of evolution, and innovations like artificial intelligence are accelerating digital transformation even further. The fourth industrial revolution, which is transforming the project management industry and disrupting virtually every aspect of business, was sparked by advanced technical advancement. Organizations utilize project management as a strategy to implement and oversee the knowledge, expertise, competencies, and experience necessary to complete projects. A project manager takes charge of a certain goal and oversees team members and other decision-makers to guarantee timely and successful project completion. Due to the fourth industrial revolution and a rapid decline of traditional project management techniques, the most recent wave of project management is now upon us. Professionals working on projects must embrace cutting-edge innovations and have a thorough awareness of how technology is affecting the project profession in today's digitally driven environment.

Moreover, IR paved the way for digital transformation by bringing a revolution in project management. Therefore, managing digital transformation with project management is a constantly evolving responsibility assigned to industry professionals. When employed properly, technology may enhance agile teams, maximize employee well-being, and support the implementation of better organizational processes and procedures. Furthermore, digital transformation has created cutthroat competition for organizations globally, and their survival significantly depends upon the following five principles:

Principle of Leadership – For project managers to get the most out of IR 4.0, it is important to maintain a positive attitude and recognize that digital transformation encompasses more than just computing power. Customer requirements are growing, and elevated standards must be met, as new technologies enhance the customer experience. Soft skills are more crucial than ever for project managers, but social skills will also be more valued across the board in the workforce. To advance in a market driven by digital technology, adaptability, versatility, and the ability to foster innovation are necessary.

Principle of Automation – Instead of being afraid of automation because of its possible effects on the workforce, come up with ways to extract more from project teams than simple clerical work. Your team is free to focus on developing their interests and abilities, which will eventually bring value to the organization, as computers handle more tedious and mundane work. Teams made up of people with different backgrounds and skills will increasingly be the norm, and project management and leadership responsibilities will probably cross over more in the coming years.

Principle of Emotional Intelligence – To successfully balance innovation and disruption with the consistency of organizational processes, leaders will need to implement 4IR. Leadership will increasingly revolve around natural collaborative processes as a growing number of businesses adopt advanced robotics and artificial intelligence. To counteract organizational policies, this new dynamic will frequently call for project management and leadership responsibilities to overlap. As support engagement, motivation, and general zeal throughout organizations are important, the value of acquaintances and emotional intelligence are going to be crucial for project managers.

Principle of Agility – Leaders need to get over the idea that digital transformation is only about faster computers. The first computer transformation is over, and the second, which is transforming technological capabilities, organizational resources, and consumer interactions, is well underway. Although the digital transformation has brought about significant changes, the essential elements of leadership have not changed. To effectively operate within 4IR, professionals must rely on guiding concepts. In addition to acting as a catalyst for transformation and encouraging multidisciplinary teams across the organization, the adaptive project manager will play a bigger part in recognizing value, encouraging creativity, and promoting flexibility.

Principle of Readiness – It takes knowledge and visualization to envision new customer experiences, goods, and services to embrace digital transformation and enhance leadership. Understanding the characteristics of the business culture, embracing digital flexibility, and getting rid of ineffective bottlenecks are all important. Now, however, project managers must apply these competencies in a complicated, artificially intelligent manner.

CHAPTER 2

Concept of Industry 4.0

2.1 INDUSTRY 4.0 OR IR 4.0

Industry 4.0, or the fourth industrial revolution can be defined as the Internet of Things (IoT) and Internet services combined into the production environment, in which all industry companies globally connect and manage their devices, factories, and smart warehouses online. It incorporates physical systems through the exchange of information and data that triggers actions (Gilchrist, 2016; Ghobakhloo, 2018). The revolution of IR 4.0 has compelled the transformation in the construction industry by exhibiting the power of digital construction with digitally available data, and the capabilities of digital technologies and the internet could be used to automate tasks and generate valuable electronic information (Alaloul et al., 2018). By implementing technologies across all facets of their business processes, companies can undergo a digital revolution through IR 4.0 and effect significant change. Being one step ahead of the competition in the digital world has benefits for businesses, including increased agility, productivity, and, ultimately, the development of new value for consumers, staff members, and shareholders. Digital transformation forces a shift from traditional thinking to a more interactive and innovative strategy for most enterprises. These innovative methods of working produce novel concepts that, in turn, can raise consumer happiness, foster employee innovation, and advance basic business performance. This digital transformation may include the utilization of robotics, IoT, artificial intelligence (AI), augmented and virtual reality (AVR),

 DOI: 10.1201/9781032621760-2

Cloud Computing, Big Data, and Building Information Modeling (BIM) in its day-to-day business activities.

BIM is defined by ISO 19650-1:2018 as the use of a collaborative digital model of a created asset to facilitate decision-making and accelerate the design, construction, and operation phases (BSI, 2019). The incorporation of BIM into the IT environment allows for the transition from the present practice of "event response" to the practice of "event prediction" (Woodhead et al., 2018). By integrating BIM into Cloud Computing, project stakeholders can work together in real-time from different locations to improve decision-making and confirm project viability (Craveiroa et al., 2019). Together with BIM, IoT can maximize productivity (Li et al., 2016a), improve the flow of information throughout the project life cycle (Dave et al., 2015), maximize energy proficiency (Bottaccioli et al., 2017), and enhance safety and security (Fang et al., 2016, Li et al., 2016b), as well as resource planning, management, monitoring, and control (Fang et al., 2016).

Consequently, Industry 4.0 was the fourth conceptualized revolution after Industry 1.0, Industry 2.0, and Industry 3.0 revolutions, as shown in Figure 2.1, in the manufacturing industry initially, but this view has changed in recent years (Xu et al., 2018). Industry 4.0 is presently being digitally updated in all industrial and consumer markets, from smart launches to digital integration to all pricing channels (Schroeder et al., 2019). Experts, government officials, and technicians regularly employ Industry 4.0 technologies for digitalization and industrialization, distribution channels, and value chain industries (Liao et al., 2017). Previous scholars frequently clarified the notion of IR 4.0 by proposing fundamental principles and advancements in technology, to enhance comprehension of this concept (Zheng et al., 2018). These design principles provide manufacturers with a "how-to" experience in assessing Industry 4.0 compliance and implementing the processes and solutions needed to change the process. The development of Industry 4.0, which concentrates only on the latest digital technologies, is known as the digital industry.

Figure 2.1 illustrates the transformation of different industrial revolutions, that is, from IR 1.0 to IR 4.0. IR 1.0 began in the eighteenth century with the use of steam power and the mechanization of industry. The simple spinning wheels that had previously been employed to make threads were replaced by a mechanized version that produced

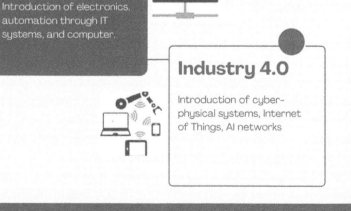

FIGURE 2.1   Industrial Revolutions. Shows a Transformative Journey from IR 1.0 to IR 4.0.

eight times the volume at the same time. Steam's strength was well-known. The use of it for industrial purposes was the most significant innovation for increasing human productivity. Alternative to using human labor, steam engines could potentially be used to power weaving looms. Likewise, in the nineteenth century, the development of electricity and the assembly line symbolized the beginning of IR 2.0. Henry Ford (1863–1947) got the concept for mass production at an abattoir in Chicago where pigs were hung from conveyor belts, and every butcher completed only a part of the slaughtering procedure. By implementing these concepts in the production of automobiles, Henry Ford completely transformed the sector. A single station used to be used to assemble a whole automobile; now, small batches of vehicles are manufactured on a conveyor belt, which is less time-consuming and inexpensive.

Beginning in the 1970s, IR 3.0 brought another revolution with the utilization of computers and memory-programmable controls. As a result of the advancement of these technologies, companies were able to fully automate the manufacturing procedure without requiring the involvement of humans. Unaided robots that follow preprogrammed instructions are prominent instances of IR 4.0, which are now being utilized commonly. IR 4.0 has been expanded on the technologies of IR 3.0 and is characterized by using information and communication technologies. A network link enables the expansion of computer-based manufacturing systems and, in a sense, the creation of an online digital twin. These facilitate interactions with other facilities and the exchange of data. This is the subsequent step of manufacturing automation. The interconnection of all systems results in "Cyber-Physical Production Systems (CPS)" and, as a result, "smart factories," where individuals, elements, and production processes communicate across a network and manufacture essentially on their own.

2.2 REASONS TO ADOPT INDUSTRY 4.0

The concept of IR 4.0 concept encompasses efforts to create integrated wireless devices that work seamlessly across different digital devices, technologies, and programming languages (Xu and Duan, 2019). As an extraordinary advancement, the German government began utilization of the Industry 4.0 concept in 2011 to electrify the production of trains (Newman et al., 2020). In this regard, Industry

4.0 applications have emerged extensively and are now widespread in existing production systems. These applications comprise cloud storage (Geng and Liu, 2016), the ambiguous structure of the theory (Sisinni et al., 2018), the design and construction of steel structures, and the installation of fast product assemblage. Digital applications and solutions can better visualize the entire product and then make product-independent decisions (Sony and Naik, 2019). This change in industrial processes has enabled the stakeholders to collect and analyze data in real-time, resulting in faster, better, and more efficient production and more flexible work practices (Flynn et al., 2017). These improvements are the advantages of direct machine-to-machine communication. Thanks to Industry 4.0's technological advances, production continued to compete more financially (Zhong et al., 2017). Industry 4.0 incorporates a combination of advanced and modern technologies, comprising innovations in AI (Sony and Naik, 2019), Big Data which exhibits large volumes of data (Trotta and Garengo, 2018), IoT which is a network of various physical systems (Trappey et al., 2017), Sensor-Based Technology (Sormaz and Malik, 2018), Press 3B, Cloud Computing (Craveiroa et al., 2019), Advanced Software Utilization, that is, BIM (Bensalah et al., 2019; Sheikhkhoshkar et al., 2019), and Cybersecurity (Fisher et al., 2018). It is a highly efficient network-physical environment that enables machines to communicate and interact directly with each other without human involvement through direct communication and decision-making capability optimization in production environments (Sormaz and Malik, 2018). The key benefits of Industry 4.0 adoption are provided in Figure 2.2.

Figure 2.2 has shown the benefits of adopting IR 4.0, such as the manufacturers and suppliers will be able to reduce their logistic times from their facility to the markets, clients will have the facility to respond to the vendors for making products more efficient and user-friendly, mass production will be possible with minimized production costs due to digital transformation, work environments will be improved with safer and work-efficient environments, the resource utilization will be optimized, and coordination and communication among concerned stakeholders and clients will be enhanced.

Businesses that use the Industry 4.0 methodology operate digitally, utilizing Big Data, Analytics, AI, and Cloud Technology to collaborate with connected equipment and intelligent schedules. As a result,

Reasons for Industry 4.0 adoption

1 SHORTER TIME TO MARKET
A shorter time-to-market for the new products.
This will enhance the availability and accessibility for end-users.

2 CLIENT RESPONSIVENESS
Enhancement in client's satisfaction.
Improvement in business ability to respond to the client's inquiries and their fulfilment in time.

3 MASS PRODUCTION WITH REDUCED PRODUCTION COSTS
Enable the mass productions without considerably impacting the overall production costs.

4 ENHANCED WORK ENVIRONMENTS
Work environments can be enhanced by the incorporation of flexible and friendlier work practices.

5 EFFICIENT RESOURCE UTILIZATION
With declining non-renewable resources, industry 4.0 will assist in more effective and efficient utilization of natural resources and energy.

6 ENHANCED COORDINATION AND COMMUNICATIONS
Considerable improvement in coordination and communications between project teams. Reduced conflicts and disputes due to better communication in the projects.

FIGURE 2.2 Industry 4.0 Adoption Benefits. Shows Key Benefits for Adopting the IR 4.0 Strategies.

the operation is responsive and has total command of its demand, inventory, and manufacturing processes. Other reasons to adopt IR 4.0 include the following:

1. Highly innovative

Using 3D design capabilities, experiments, and prototyping, as shown in Figure 2.3, may be carried out quickly and even virtually. Better decisions result from including consumers, suppliers, and staff throughout the entire design and production process.

2. Increased productivity

Due to its ability to manufacture a range of commodities at higher rates and of better quality, to gain from enhanced supply chains and distribution frameworks, as well as from more rapid decision-making throughout the board, the smart factory can function at a much higher level of productivity. A cyber-physical system that analyzes data, powers machine-learning algorithms, and continuously learns is referred to as a "smart factory." All the processes in the smart factory are interconnected through Cloud Computing and work together in a synchronized way, as shown in Figure 2.4. In other words, all the key segments working in a company such as suppliers, sales management

FIGURE 2.3 3D Design Capabilities, Experiments, and Prototyping.

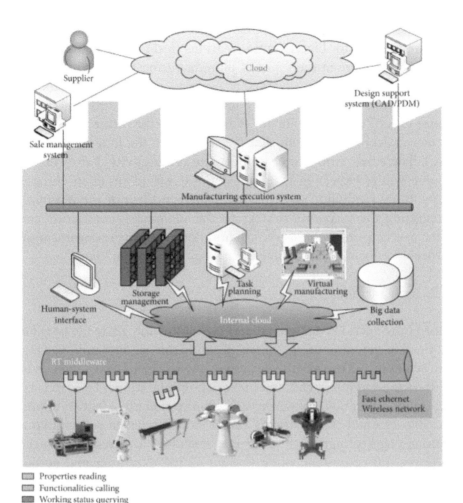

Supplier

Cloud

Design support
system (CAD/PDM)

Sale management
system

Manufacturing execution system

Human-system
interface

Storage
management

Task
planning

Virtual
manufacturing

Internal cloud

Big data
collection

RT middleware

Fast ethernet
Wireless network

 Properties reading
Functionalities calling
Working status querying

FIGURE 2.4   Smart Factory.

teams, human systems, storage management, task planning, virtual planning, manufacturing execution systems, and robots employed on assembly lines are all digitally connected, transferring and sharing data among them for efficient business operations.

Manufacturers must first understand that they must start making the shift to Industry 4.0 immediately. More than 90% of makers of consumer goods consider digital transformation to be a major priority, according to the Manufacturing Report of 2020 (Manufacturer, 2020). What exactly does it suggest considering Industry 4.0? The immediate transition to significant futurization, as certain business leaders fear, is not necessary. It refers to accepting the actions we can take right now to pave the way for bigger transformations in the future. Automation will be used more in the future if manufacturing facilities utilize the potential of automation, advanced technology, and connectivity today.

Integrating operational and IT infrastructure presently will give a better idea of what the company will require in the future, allowing it to benefit from innovations that are tailored to the long-term objectives. There are various ways to evaluate productivity. For example, maybe you're expecting that Industry 4.0 would result in cheaper costs, more flexibility, or better quality. Regardless of the response, integrating the corporate strategy to the advantages connection can offer can help make Industry 4.0 advantageous to the organization rather than focusing on digital technology for its own sake.

3. Agile methods

Shorter production runs and greater customization are made possible by flexible manufacturing, which expands the product mix, variety, and scalability. Organizations can work with clients and suppliers to address their requirements and make accommodations. The capacity, flexibility, and ability of an organization to quickly adapt to unpredictable external circumstances and terrain is referred to as (organizational) agility. Significant external difficulties are presented by the Industry 4.0 ecosystem, often known as the execution and acceptance of I4.0 technologies for establishing an edge over competitors.

It is also asserted that I4.0 technologies are naturally agile. The organizations will noticeably increase their supply chain, production, personnel, etc. agility. Other researchers assert that the first organizations to adopt and execute I4.0 technologies and take full advantage of the competitive benefits offered by the Industry 4.0 ecosystem would be those who are nimble and rapidly and easily react to external developments. These studies highlight the highly intricate nature of the Industry 4.0 ecosystem, which necessitates a significant reorganization of organizational culture, infrastructure, and leadership. In terms of changing the organization's culture, leadership, and facilities, establishing the Industry 4.0 ecosystem would be difficult. A company with agility as a (dynamic) competency will be able to swiftly reconfigure and create a seamless process for deploying I4.0 technology.

4. Lower costs

In Industry 4.0, high levels of automation lead to less labor-intensive tasks, fewer resources being wasted waste, and more effective operations, all of which directly decrease operating expenses. The use of technological tools enables everyday activities to be optimized, which increases the productivity of manufacturing and administrative procedures. Industry 4.0 will provide businesses with highly sophisticated efficient operations while employing innovative and powerful technical resources. Practically speaking, businesses have the chance to optimize resources for improved outcomes that boost competitive edge. Here are a few instances where using an effective smart factory can result in cost savings in this Industry 4.0 execution:

a. Decrease in unforeseen production shutdowns

One of the major industry restraints is the issue with equipment causing unscheduled production breakdowns, which causes production interruptions and necessitates the need for specialists and professionals for emergency needs. The so-called preventative maintenance, which constantly assesses the equipment to indicate breakdowns, can advance with the adoption of Industry 4.0. Utilizing sensors that are mounted on the equipment can send data to the cloud via systems, for maintenance tasks. In contrast to preventive maintenance, which requires part repair or substitution even when the machine does not require it, it is possible to determine the requirement for intervention in advance, minimizing equipment repairs and inspecting time.

b. The enhanced atmosphere for employment

Big Data capabilities that regulate the environment's humidity, temperature, assets, and other plant-related information help optimize the working environment. A more productive atmosphere with air-conditioning, among other things, leads to increased workplace happiness, which increases productivity and profitability for the organization.

c. Team members engaging in more tasks that add value

Because of the integration of the controlling machinery and its enhanced autonomy, factory operations that were traditionally handled by workers are now almost fully accomplished by machines. This makes it feasible to hire specialists to complete more complex and strategic jobs that are specifically meant to create results. Upgrades will be necessary regularly when new tasks are added to fulfill Industry 4.0 requirements.

5. Improved workplace

Technology has been incorporated into their operational processes to recognize and encourage their staff. Workplaces with ergonomic workplaces, better training, and open communication lead to safer, more satisfying jobs that support career advancement within the organization. The future of work will be impacted by Industry 4.0 and hybrid technologies, which offer the interconnection required to improve the work environment and corporate culture. According to what humans say, the only thing that is consistent in life is transformation. The COVID-19 epidemic fundamentally changed how today's work environment operates, as evidenced by trends that point to an expanding need for hybrid workplace models. Nowadays, a lot of businesses use hybrid workforce models to run their businesses. They are, therefore, implementing the technologies required to support this kind of adaptable approach.

The concept of "Industry 4.0," or the subsequent evolution of the way individuals conduct themselves at work and home because of technological advancement, is changing how people view the nature of employment. Interconnectivity is what this revolution is all about. The Internet of Things (IoT), smart structures, devices such as sensors, AI-driven data, and Big Data are examples of technology that has resulted from Industry 4.0. The modern workplace that we know today is the result of these technologies. Industry 4.0 is revolutionary in its way, much like the previous industrial revolutions were. The preceding industrial revolution, known as Industry 3.0, began with the development of mainframe computing and, subsequently, automation. As a result, hardware, software, and other fundamental technologies were finally created, igniting Industry 4.0.

Companies must adopt cutting-edge technologies that will give them an advantage if they want to succeed in the future. Many businesses now encourage employees to split their time between the office and the home. It's referred to as a hybrid workplace. Businesses are now more able to count on technology than they did before the pandemic because of Industry 4.0 and the hybrid workplace. Hybrid work environments and Industry 4.0 place a strong emphasis on the usage of Integrated Workplace Management Platforms (IWMS), such as Office Space, which includes capabilities like desk booking, arrivals, and more. Due to the advancement of Industry 4.0 and IoT, these add-ons enable the property and business sectors to interact.

EXAMPLE 2.1 THOUGHT BOX FOR STUDENTS

It's crucial for businesses to adopt cutting-edge and innovative technologies that will give them an edge if they want to succeed in the future.

For instance, *"Facebook announced Horizon Workrooms,"* which enables remote workers to join a virtual office using Oculus VR headsets. The "Metaverse," a virtual world where individuals utilize avatars to socialize, conduct business, and amuse themselves, is depicted as including users' avatars. Although we can't anticipate exactly which technology the future will bring, we do acknowledge that organizations are adopting "hybrid" and that remote work is on the rise, so this idea isn't too far off from being reality.

---

6. Increased resource efficiency

Through IR 4.0 technologies, resource efficiency can be obtained at all levels within an organization. Resource optimization capabilities are directly linked with data transformation levels and data flow processes, as shown in Figure 2.5. Moving to higher levels of data transformation levels will result in optimized human decision-making. With the consistent transfer and sharing of data through data flow channels, that is, data collection, data integration, and data analysis, resource optimization can considerably be enhanced.

2.3 STRATEGIES IN IR 4.0

The integration of machine learning and cognitive power enables computers to handle extremely complicated jobs, but they cannot substitute for deep expertise. Rather, they appear to be more effective than people at undertaking repetitive activities. Significant changes are taking place in the way the global production and supply network operates

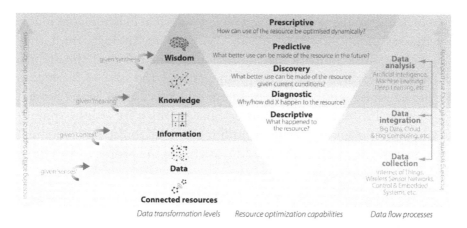

FIGURE 2.5   Resource Optimization Framework.

because of the ongoing automation of traditional manufacturing and industrial practices, the application of cutting-edge smart technology, substantial machine-to-machine communication (M2M), and the Internet of Things (IoT). The benefits of this integration involve expanding automation, enhanced communication and self- tracking, and the introduction of intelligent gadgets that can analyze and identify issues without involving humans. This section will discuss in detail the commercially available technologies and research or development phases.

2.3.1 In-Use Technologies of IR 4.0

In line with the design strategies, digital transformation in Industry 4.0 is based on the implementation and integration of the latest advanced information processing technology, such as sensors, self-driving automotives, robotics, augmented and virtual reality (AVR), data analytics, Cloud Computing, Internet of Services (IoS), Design and Computer-based Construction on High-Performance Computing (HPC-CADM), and AI (Chen et al., 2017; Hofmann and Rüsch, 2017). These technologies, along with IoT, BIM, Smart Factories, Big Data, 3D Printing, Modular Construction, and Nanotech, are being used globally.

Furthermore, the digital transition in Industry 4.0 includes the registration and integration of all standard components. The Digital Supply Network (DSN) provides direct connections to state-of-the-art devices, so that smart providers, power tools, smart companies, technology, and all services that connect to smart include global information for smart communication devices (Ardito et al., 2019). This integration component is supported by IOS and the Human Network (IOP), which connects consumers and products through DSNs and directs Product Development to Services (PAO) and business development (Ghobakhloo, 2018). The use of such smart factories, CPPS, IIOT, Big Data, and Cloud Data form the basis of products that can run on machines, human devices, equipment, and administrators, and real-time effects, such as conferences (Chen et al., 2017). Communication, transparency, control, and supervision of production have managed to reduce the risks related to waste, energy, and leadership (Ghobakhloo, 2018; Schroeder et al., 2019). All smart companies and good networks provide valuable opportunities to evaluate and improve performance (Leng et al., 2019). For example, a development device can model global devices and test digital products over their lifetime. Access has been adapted to the changing needs of individuals (Niaki et al., 2019). Introducing IOP, as well as meeting current client needs and priorities, comparing consumer behavior understanding business, and assisting good business for smart production centers, helps in achieving key objectives (Torn and Vaneker, 2019). Figure 2.6 provides an overview of IR 4.0 strategies that contributed to the industrial theory of IR 4.0 and lists the core industry strategies and scientific methods of IR 4.0 found in the literature.

2.3.2 IR 4.0 Technologies That Are Not Fully Available for Use

Currently, lots of research is being carried out on the advanced IR 4.0 technologies, and yet not fully available for commercial use. These include the following:

2.3.2.1 Industrial Internet of Things (IIoT) or Network Physical Production Systems (CPPS)

The latest technology in Industry 4.0, that is, IIoT or CPPS, is not readily available in the latest technology products (Ghobakhloo, 2018). This sophisticated technology is based on the implementation and integration of different Identity of Things (IDOT) components into a quality network. For example, IIOT and CPPS as integrated and interconnected memory, based on AI integration, machine-to-machine communication, industry managers, smart key, cloud data, large data analytics, and semantic technology establish strong network controls. It is a system that ensures the efficiency and reliability of the production process (Sisinni et al., 2018).

FIGURE 2.6   Strategies of Industry 4.0.

2.3.2.2 Quantum Computing Adoption in Manufacturing

Through the resolution of challenging optimization issues and the improvement of simulations, quantum computing can revolutionize manufacturing procedures. Large-scale execution, however, have not yet been realized, and real-world applications continue to be in the initial phases of the study.

2.3.2.3 AI using Neuromorphic Computing

Neuromorphic computing aims to mimic the organization and function of the human brain in artificial neural networks. While it has a lot of potential for cutting-edge AI and machine learning tasks, research and development on it are still ongoing.

2.3.2.4 Swarm Robotics

In swarm robotics, many small robots are coordinated to function as a single cohesive entity. Numerous uses for these swarms might be found in the industrial, logistics, and other sectors of the economy. Even though there have been a few experimental explanations, they have not yet become generally accessible for commercial application.

2.3.2.5 Advanced Integration of Human-Robots

Research has been done in this area to make collaboration between people and robots more secure, more rational, and more successful. This entails creating methods for shared systems of control, motion detection, and natural language interface.

2.3.2.6 Smart and Self-Healing Materials

Self-healing materials, which can fix damage on their own, and smart materials, which can sense and react to environmental alterations, are still in the development and research phases for a variety of industrial uses.

2.3.2.7 Solutions for Energy Harvesting

Multiple energy harvesting techniques that could operate IoT sensors and devices in Industry 4.0 contexts remain the subject of research. Vibration energy mining, thermal energy processing, and RF extraction of energy comprise some of these techniques.

2.3.2.8 Utilization of Blockchain Technologies for Supply Chain Management

Blockchain-based technology is being investigated for applications in supply chains to provide a means of tracking and transparency, but integration and mainstream adoption are still developing.

The limitations of the IR 4.0 integration are limitless, as these terms contribute to the IR 4.0 integration (Li et al., 2016a). Collaboration ensures that the various aspects of quality communication, such as management systems, intelligence tools, firms, smart devices and products, connected customers, decision-making processes, and human resources, can connect, communicate, and coherently exchange data (Zheng et al., 2018; Zhong et al., 2017).

EXAMPLE 2.2 THOUGHT BOX FOR BUSINESSES

Organizations need to create and put into practice effective strategies if they want to flourish in this disruptive time. The businesses put into practice several crucial tactics to address these issues and take advantage of Industry 4.0's advantages.

1. Application of smart manufacturing.
2. Making decisions based on data.
3. Transformation of the workforce and skill development.
4. Partnerships and cooperation.

2.4 INDUSTRY 4.0 IMPACTS ON TECHNOLOGY

The impact of Industry 4.0 on technology and society is diverse. Traditional tools, techniques, and technologies have become obsolete because of the Industry 4.0 revolution. The approach to conducting the projects by the senior management will be transformed as per the needs and requirements of Industry 4.0. To understand how business leaders are juggling the shift to Industry 4.0 with cutting-edge technology that will assist their organizations to flourish in their sectors while acting more ethically, Deloitte (Renjen, 2020) interviewed more than 2,000 corporate executives from 19 different countries. The development and implementation of an integrated strategy for the use of Industrial 4.0 technology, from talent acquisition and training to global protection, should be the focus of all organizations around the world. The results of this survey are presented in Figure 2.7 and indicate that IoT, AI, and Cloud Computing will have the maximum impact (72%, 68%, and 64%) on technology adoption. The result of the survey indicates that the respondents termed the utilization of IoT, AI, and Cloud Computing as the most important digital transformative tools. The adoption of these tools will trend in all industries globally, that is, manufacturing, textile, automobile, space, construction, medical, aeronautical, etc. The results of Figure 2.7 should not be misinterpreted in a way that other mentioned

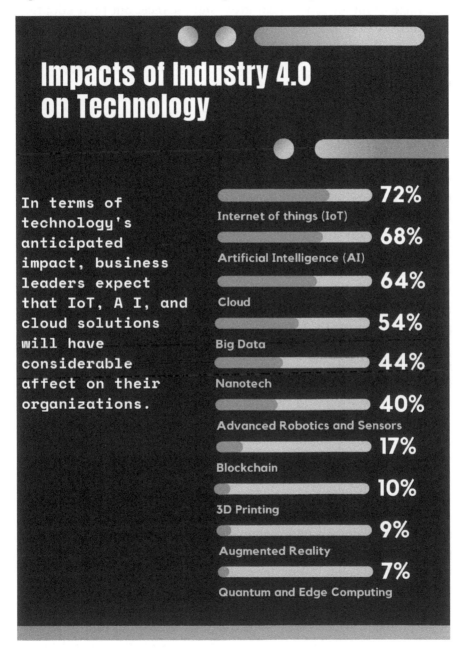

FIGURE 2.7   Industry 4.0 Impacts on Technology.

technologies have no significance. The other technologies mentioned in Figure 2.7 such as Big Data, Nanotech, Advanced Robotics and Sensors, Blockchains, 3D Printing, Augmented Reality, Quantum, and Edge Computing also bear great significance, but their adoption currently is slow due to initial costs, regulatory and legal requirements, availability of local and international standards, and safety protocols. As soon as these hurdles are removed, the adoption of these technologies will be made easy.

2.5 INDUSTRY 4.0 IMPACTS ON SOCIETY

To improve business productivity and competitiveness, Industry 4.0 is typified by the integration of various digitally based technologies into production procedures. It is the foundation for the "smart factory" and a new organization of efficient operations that is favorable to advances in efficiency and to a more effective and more environmentally friendly allocation of resources. It relies on the connectedness of objects and cyber-physical systems. Industry 4.0 undoubtedly has enormous implications for economic growth and society, but its implementation has also sparked worries about its potential adverse effects on both. The subsequent sections will discuss the negative and positive impacts of IR 4.0 on society.

2.5.1 Negative Impacts

Industry 4.0–driven integrated advancements are defining a new era of globalization and raising a lot of concerns about how these developments will affect society. Some could contend that the greatest threat facing humanity now is globalization, particularly in the age of Industry 4.0. Its negative repercussions include an increase in income disparity, harmful contamination of our environment, and the replacement of human labor by robots. However, globalization is often mistaken for globalism, an ideology that supports a neoliberal global order over national interest. The actual source of unhappiness, annoyance, and insecurity, particularly in Western cultures, is not globalization, but rather globalism, an ideology that supports a neoliberal global order at the expense of national interests.

Like this, income inequality has its roots in our institutions and system of government, which gives favored capital owners an ever-increasing percentage of GDP. Although the impact of innovation indeed causes industrial revolutions to frequently alter the structure of the labor market. However, this shouldn't result in widespread unemployment. According to history, new inventions frequently lead to an increase in employment prospects. Today, many of us have positions that did not exist when we were teenagers. The jobs that will be disrupted by new technology and the immediate effects that will have on people's lives and their communities, rather than a spike in unemployment brought on by Industry 4.0, are what we feel people are most concerned about. The labor market, however, has become more divisive. Income disparity is largely caused by a trained elite set of people who can use technology to be more productive, replace other people's labor, and are compensated accordingly. However, these negative consequences are only transitory and small-scale, but ineffective leadership from the government, business, and academic sectors can have a considerable negative influence.

2.5.2 Positive Impacts

2.5.2.1 Enhanced Labor Markets

In fact, by proactively addressing the market's talent gap, employment polarization can be prevented. It is feasible to determine which tasks and jobs are amenable to automation and hence run the risk of being superfluous by analyzing them. On the other hand, if the workforce's particular skills are also examined, it should be feasible to determine how these employees may excel in other ways and then provide them with education in those areas. Innovative technologies, as well as improved regulations coupled with more sustainable business models, can lead to new possibilities for employment and income. For the outcomes to become more equitable for every member of society, we must all work together on a greater scale. Only if the private sector, academia, and government – starting at the local level, moving up to the national level, and then internationally – can adapt to change and work together towards a common greater purpose can some of the most serious concerns of our day be resolved.

Because automated technology produces better results and sufficient progress, technological advancements are a threat to the employment market. The labor market today, however, benefits from accepting change and modifying one's skills to meet the demands of a changing society despite the so-called "threat" that it poses. The future of "work," as shown in Figure 2.8, will promote sustainability and advancement while providing unmatched prospects. A more technologically based, highly skilled labor pool is being forced onto the labor market because of the transformation, and this has the potential to offer up opportunities for disadvantaged people in the process. The work organization is changing in the meantime to reflect the demands of the market. Technological advancement in automation is transforming the current workplace in addition to the labor market by replacing mechanical, muscle-added, monotonous duties with programmable, self-enhancing algorithm information.

2.5.2.2 Role of Governments and International Organizations

Infrastructure, education, regulation, and government would all need to undergo significant adjustments. This will undoubtedly need

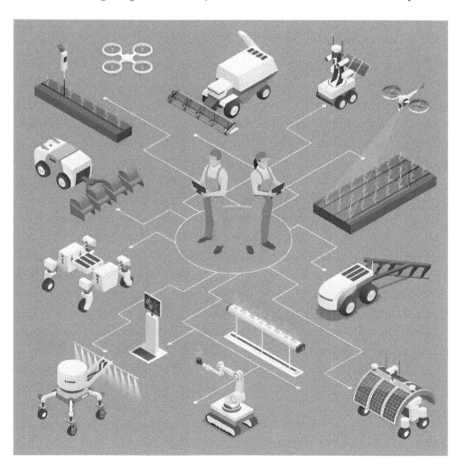

FIGURE 2.8 Future of Work.

bravery, strong leadership, and entrepreneurialism from all parties involved and necessitate cooperation at all levels of geography. The choices we make now will ultimately determine how the future turns out and its impacts on society. Thus, discussions about the Industry 4.0 paradigm are still going on. Most economists concur on its technological features; however, there is disagreement over its social and economic results. Some argue that it emerged as a result of a decline in population in Germany and Japan, prompting industrial companies to employ a greater amount of automation, robotics, and state-of-the-art digital technologies. Others believe that Industry 4.0 will enable industrialized nations to move their manufacturing operations to areas with a shortage of labor. However, Industry 4.0's technological advancements should focus on sustainability and climate change mitigation.

2.5.2.3 Impact on Climate Change and Sustainability Adoption

There is an urgent need to implement significant changes cooperatively given the worldwide impact of climate change (global warming) and the difficulty of combating its impacts in a non-cooperative fashion. There are international policy structures that allow interdisciplinary efforts. Two important worldwide policy frameworks for sustainable development that specifically include adjustments to industry to achieve sustainability are the Paris Agreement and the United Nations Sustainable Development Goals (SDGs). Furthermore, the G20 summit that took place in Germany in 2017 specifically addressed this problem. Our economies and society have a great deal of potential to change and realign because of Industry 4.0. The Industry 4.0 paradigm might assist the global community in implementing the necessary technological advancements and accelerating the trend toward decarbonization. Deloitte's (Renjen, 2020) survey on the anticipated impacts of IR 4.0 on society, as shown in Figure 2.9, also supports this notion that the issue of climate change can be managed effectively through the incorporation of sustainability. The survey indicates that 89% of businesses will be significantly impacted by climate change. This is a huge figure and should be focused on with mutual consensus among stakeholders globally. Moreover, it is also very motivating that 91% of the stakeholders engaged in the survey are directly or indirectly adopting sustainable practices in their business operations.

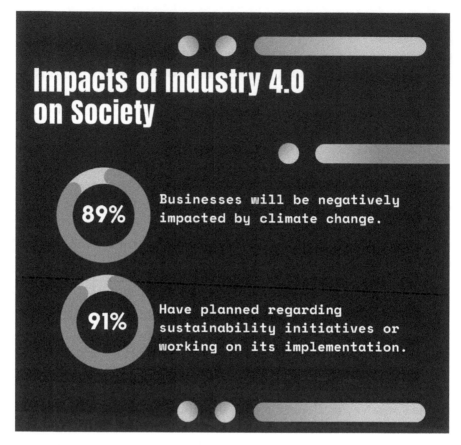

FIGURE 2.9   Industry 4.0 Impacts on Society.

2.6 INDUSTRY 4.0 IMPACTS ON THE ENVIRONMENT

Production facilities must integrate hundreds of thousands of IoT devices with their equipment to monitor and evaluate processes. However, because of the enormous energy demand brought on by the rapid flood of electronic devices that fill a facility, the electrical grid systems may end up under more strain. Even though many IoT devices are battery-powered, large data transmission requires an enormous amount of computer processing power. Adopting Industry 4.0 has some problems, one of which is that it encourages profits more than it contributes to the exhaustion of natural resources. Big manufacturers must see the need to become more environmentally friendly and refrain from encouraging excessive consumption. Other issues with Industry 4.0's environmental effects include the following:

   a. Destruction of forests and other forms of raw resource exhaustion.

b. Difficulties with human health.

c. Contaminated water in the ground.

d. Disruption of natural processes.

e. Huge data centers' waste and pollution.

Increasing the usage of more environmentally friendly alternative energy sources like solar and wind is one element of the solution to reduce CO_2 emissions in the atmosphere. The use of renewable energy sources by utilities is already contributing to lower electricity prices. Sustainability is becoming a more important component of business objectives. Industry 4.0 is also assisting companies in becoming more environmentally friendly. Switching to clean, renewable energy is a necessary step toward achieving environmental goals. A comprehensive framework is provided in Figure 2.10 that illustrates the integration of IR 4.0 with sustainability to attain Sustainable Environment 4.0 or Sustainability 4.0. It means that the only way to move from traditional sustainability to Sustainability 4.0 lies with the incorporation of IR 4.0 technologies irrespective of the industry.

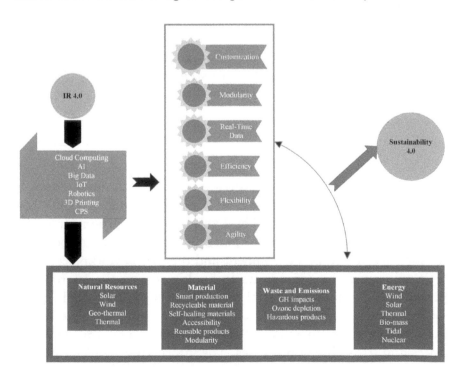

FIGURE 2.10 Sustainability 4.0 Framework.

2.7 INDUSTRY 4.0 IMPACTS ON THE WORKPLACE

Leveraging technology, in the long run, is essential for businesses to develop, since people still want to spend some time in the office. The days of working from nine to five, all day in the office, are over. Nowadays, there are more options and opportunities for employees in terms of where and how they perform their duties. The technology that offers the most information regarding how to prepare for the future of work must be available to companies. Because the workplace is constantly evolving, the new hybrid environment will benefit from insights derived from data. The epidemic expedited the implementation of Industry 4.0. It accelerated the integration of technology into the workplace to increase productivity, reduce expenses, and enhance the overall working environment.

EXAMPLE 2.3 THOUGHT BOX FOR STUDENTS

The innovations brought about by Industry 4.0 are significantly shaped and enhanced by students. Here are a few strategies students can use to embrace and take advantage of the opportunities provided by Industry 4.0:

1. Gaining digital literacy.
2. Accepting ongoing learning.
3. Taking part in projects and competitions related to Industry 4.0.
4. Working in partnership with business and academia.
5. Encouraging innovation and business ventures.
6. Promoting ethics and raising awareness.
7. Taking part in various social and community-related activities.

To improve the employee experience, businesses must adopt a more flexible work environment that combines Industry 4.0 and a hybrid work environment. The hybrid work environment is the workplace of the future. And to preserve interconnectivity between the conventional work environment and the digital workplace, businesses must use the appropriate technologies. Use the appropriate technologies because it's crucial to embrace the correct tools, from collaboration platforms like Slack and video conferencing solutions like Zoom or Teams to an IWMS that may assist in organizing workspaces and streamlining processes. The information needed to make better decisions must be gathered, particularly regarding the places and methods of employment. Therefore, businesses must introduce technology that supports a hybrid philosophy.

EXAMPLE 2.4 THOUGHT BOX FOR BUSINESSES

Facility managers (FMs) may optimize office operations with the aid of facility management software such as Integrated Workplace Management Systems (IWMS) and Integrated Facilities Management (IFM).

IFM software aids FMs in comprehending events occurring on various sites and building stories. It focuses on uniformly streamlining operations and business processes. IWMS software links independent resources to create a single, completely integrated workplace environment.

REFERENCES

Alaloul, W. S., Liew, M. S., Zawawi, N. A. W. A. & Mohammed, B. S. 2018. Industry revolution IR 4.0: Future opportunities and challenges in construction industry. *MATEC Web of Conferences*. EDP Sciences, 02010. https://doi.org/10.1051/matecconf/201820302010

Ardito, L., Petruzzelli, A. M., Panniello, U. & Garavelli, A. C. 2019. Towards Industry 4.0: Mapping digital technologies for supply chain management-marketing integration. *Business Process Management Journal*, 25, 323–346.

Bensalah, M., Elouadi, A., Mharzi, H. J. S. & Environment, S. B. 2019. Overview: The opportunity of BIM in railway. *Smart and Sustainable Built Environment*, 8(2), 103–116.

Bottaccioli, L., Aliberti, A., Ugliotti, F., Patti, E., Osello, A., Macii, E. & Acquaviva, A. 2017. Building energy modelling and monitoring by integration of IoT devices and building information models. *2017 IEEE 41st Annual Computer Software and Applications Conference (COMPSAC)*. IEEE, 914–922. https://doi.org/10.1109/COMPSAC.2017.75

BSI. 2019. *BS EN ISO 19650: Organisation and Digitisation of Information about Buildings and Civil Engineering Works, Including Building Information Modelling – Information Management Using Building Information Modelling.* London, BSI: British Standards Institution. http://www.bsi-global.com/

Chen, B., Wan, J., Shu, L., Li, P., Mukherjee, M. & Yin, B. 2017. Smart factory of Industry 4.0: Key technologies, application case, and challenges. *Journal of IEEE Access*, 6, 6505–6519.

Craveiroa, F., Duartec, J. P., Bartoloa, H. & Bartolod, P. J. 2019. Additive manufacturing as an enabling technology for digital construction: A perspective on Construction 4.0. *Journal of Sustainable Development*, 4, 6.

Dave, B., Kubler, S., Pikas, E., Holmström, J., Singh, V., Främling, K. & Koskela, L. 2015. Intelligent products: Shifting the production control logic in construction (with Lean and BIM). *Proceedings of the 23rd Annual Conference of the International Group for Lean Construction. Perth, Australia, 29–31 July 2015.* International Group for Lean Construction. https://urn.fi/URN:NBN:fi:aalto-201508164064

Fang, Y., Cho, Y. K., Zhang, S. & Perez, E. 2016. Case study of BIM and cloud–enabled real-time RFID indoor localization for construction management applications. *Journal of Construction Engineering Management*, 142, 05016003.

Fisher, L. H., Edwards, D. J., Pärn, E. A. & Aigbavboa, C. O. 2018. Building design for people with dementia: A case study of a UK care home. *Facilities*, 36(7/8), 349–368.

Flynn, J., Dance, S. & Schaefer, D. 2017. Industry 4.0 and its potential impact on employment demographics in the UK. *Advances in Transdisciplinary Engineering*, 6, 239–244.

Geng, K. & Liu, L. 2016. Research of construction and application of cloud storage in the environment of Industry 4.0. *International Conference on Industrial IoT Technologies and Applications.* Springer, 104–113. https://link.springer.com/chapter/10.1007/978-3-319-44350-8_11

Ghobakhloo, M. 2018. The future of manufacturing industry: A strategic roadmap toward Industry 4.0. *Journal of Manufacturing Technology Management*, 29(6), 910–936.

Gilchrist, A. 2016. *Industry 4.0: The Industrial Internet of Things.* Springer.

Hofmann, E. & Rüsch, M. 2017. Industry 4.0 and the current status as well as future prospects on logistics. *Computers in Industry*, 89, 23–34.

Leng, J., Zhang, H., Yan, D., Liu, Q., Chen, X. & Zhang, D. 2019. Digital twin-driven manufacturing cyber-physical system for parallel controlling of smart workshop. *Journal of Ambient Intelligence Humanized Computing*, 10, 1155–1166.

Li, C. Z., Hong, J., Xue, F., Shen, G. Q., Xu, X. & Luo, L. 2016a. SWOT analysis and Internet of things-enabled platform for prefabrication housing production in Hong Kong. *Habitat International*, 57, 74–87.

Li, C. Z., Hong, J., Xue, F., Shen, G. Q., Xu, X. & Mok, M. K. 2016b. Schedule risks in prefabrication housing production in Hong Kong: A social network analysis. *Journal of Cleaner Production*, 134, 482–494.

Liao, Y., Deschamps, F., Loures, E. D. F. R. & Ramos, L. F. P. 2017. Past, present and future of Industry 4.0-a systematic literature review and research agenda proposal. *International Journal of Production Research*, 55, 3609–3629.

Manufacturer. 2020. *Annual Manufacturing Report 2020: The Search for Stability.* Hennik Research. https://info.themanufacturer.com/amr-2020

Newman, C., Edwards, D., Martek, I., Lai, J., Thwala, W. D. & Rillie, I. 2020. Industry 4.0 deployment in the construction industry: A bibliometric literature review and UK-based case study. *Smart and Sustainable Built Environment*, 10(4), 557–580.

Niaki, M. K., Torabi, S. A. & Nonino, F. 2019. Why manufacturers adopt additive manufacturing technologies: The role of sustainability. *Journal of Cleaner Production*, 222, 381–392.

Renjen, P. 2020. *Industry 4.0: At the Intersection of Readiness and Responsibility, Deloitte Global's Annual Survey on Business's Preparedness for a Connected Era* [Online]. Deloitte. Available: https://www2. deloitte.com/global/en/insights/deloitte-review/issue-22/industry-4-0-technology-manufacturing-revolution.html?id=pk:2el:3or:4dius3 2959:5awa:6di:dr26:wef20:4ir:vanity&pkid=1006897 [Accessed 15 September 2021].

Schroeder, A., Ziaee Bigdeli, A., Galera Zarco, C. & Baines, T. 2019. Capturing the benefits of Industry 4.0: A business network perspective. *Production Planning Control*, 30, 1305–1321.

Sheikhkhoshkar, M., Rahimian, F. P., Kaveh, M. H., Hosseini, M. R. & Edwards, D. 2019. Automated planning of concrete joint layouts with 4D-BIM. *Automation in Construction*, 107, 102943.

Sisinni, E., Saifullah, A., Han, S., Jennehag, U. & Gidlund, M. 2018. Industrial internet of things: Challenges, opportunities, and directions. *IEEE Transactions on Industrial Informatics*, 14, 4724–4734.

Sony, M. & Naik, S. 2019. Key ingredients for evaluating Industry 4.0 readiness for organizations: A literature review. *Benchmarking: An International Journal*, 27(7), 2213–2232.

Sormaz, D. N. & Malik, M. 2018. Data-driven simulation modelling for progressive care units in hospitals. *Procedia Manufacturing*, 17, 819–826.

Torn, I. & Vaneker, T. H. 2019. Mass personalization with Industry 4.0 by SMEs: A concept for collaborative networks. *Procedia Manufacturing*, 28, 135–141.

Trappey, A. J., Trappey, C. V., Govindarajan, U. H., Chuang, A. C. & Sun, J. J. 2017. A review of essential standards and patent landscapes for the Internet of Things: A key enabler for Industry 4.0. *Advanced Engineering Informatics*, 33, 208–229.

Trotta, D. & Garengo, P. 2018. Industry 4.0 key research topics: A bibliometric review. *2018 7th International Conference on Industrial Technology and Management (ICITM)*. IEEE, 113–117.

Woodhead, R., Stephenson, P. & Morrey, D. 2018. Digital construction: From point solutions to IoT ecosystem. *Automation in Construction*, 93, 35–46.

Xu, L. D. & Duan, L. 2019. Big data for cyber physical systems in Industry 4.0: A survey. *Enterprise Information Systems*, 13, 148–169.

Xu, L. D., Xu, E. L. & Li, L. 2018. Industry 4.0: State of the art and future trends. *International Journal of Production Research*, 56, 2941–2962.

Zheng, P., Wang, H., Sang, Z., Zhong, R. Y., Liu, Y., Liu, C., Mubarok, K., Yu, S. & Xu, X. 2018. Smart manufacturing systems for Industry 4.0: Conceptual framework, scenarios, and future perspectives. *Frontiers of Mechanical Engineering*, 13, 137–150.

Zhong, R., Xu, X., Klotz, E. & Newman, S. 2017. Intelligent manufacturing in the context of Industry 4.0: A review. *Engineering*, 3(5), 616–630.

CHAPTER 3

Industrial Revolution and Construction Industry

3.1 MANUFACTURING AND CONSTRUCTION INDUSTRIES IN THE CONTEXT OF IR 4.0

Manufacturing is being successfully transformed into an industry that is responsive to demand, dynamic in development, and increasingly efficient in delivery due to world-leading, innovative technologies. Manufacturers all over the world are embracing the requirement to reap the benefits of the distinctive advantages of the digital age to develop this efficiency, bring down costs, and increase competitiveness in a world that is becoming more cost-conscious and competitive as we approach the fourth industrial revolution. The construction business, with one or two exceptions, is one sector that is clinging to the status quo and resistant to change. It still uses a traditional array of skills and two-dimensional processes, the bulk of which are non-digital, to manage its operations. It is quite traditional and conservative in its approach. One may argue that to increase efficiency, foster sustainability, boost safety, and decrease expensive waste, the construction sector needs to get over its resistance to innovation and grab the opportunities it presents. BIM techniques are used in some

DOI: 10.1201/9781032621760-3

areas of the industry, but the full value of BIM has not yet been fully realized because, sadly, a large portion of the industry is unaware of its true potential and is unwilling to assume the risks, both financial and operational, associated with investing in its capabilities to change the game. Businesses are making just enough progress to meet government construction mandates, but not much more. BIM encompasses much more than just digital design and realization; its complete scope should cover everything from the built asset's first installation to continuing, ongoing upkeep. Building life cycle management, to put it another way. Without the industry taking the initiative, the SMEs in the supply chains for the construction industry, which are exposed to developments in other manufacturing sectors, will either start to demand transformation or will leave the industry for other, more lucrative industries. Due to the intense competition, supply chain companies are looking to profit by collaborating with other industries and potential. If they don't, market forces may push a smaller, low-margin enterprise off the competitive map. The utilization of digital technologies to incorporate and configure data offered by the consumers and the entire supply chain spectrum can help such businesses become lucrative and effective.

If configured and integrated effectively, the considerable amount of data generated by any engineering process can be used to generate value, save costs, and significantly reduce waste. A higher return on equity will result from operations and ongoing services being driven by effective configuration management. As small businesses employ the planning capability provided to serve many customers, becoming competitive and capable of delivering value against the national and worldwide competition, that equity impact will be shared across the engineering and manufacturing industries – and beyond. That is the environment we are entering. Buildings could become distinctive and sensitive to their surroundings thanks to the creative use of digital technology, as well as intelligent, sustainable, and energy-efficient tools. By manipulating data across a 3D digital data platform, it is possible to integrate legacy buildings that have been preserved into new developments while gently balancing the old and the new and seamlessly fusing ideas from the past with those of the future. This information creates the channels that lets the client's creativity and the construction provider's dynamics flow together.

3.2 IMPLEMENTATION OF INDUSTRY REVOLUTION IN THE CONSTRUCTION INDUSTRY

New technologies are forcing their way into the construction business, which is experiencing unprecedented development. Cyber-physical infrastructure (networked control systems) and digital ecosystems are the two main fundamental components of Industry 4.0, which integrates physical and digital technology in the construction sector. The subsequent section will discuss in detail the implementation possibilities and challenges of IR 4.0 in the construction industry along with certain emerging technologies being used in the construction projects.

3.2.1 Emerging IR 4.0 Technologies for the Construction Industry

As a result of innovation and the widespread use of novel innovations that enable tasks to be completed more quickly and affordably, productivity has continuously increased across all business sectors. Contrarily, the construction industry has mostly maintained its current practices and fallen behind other sectors in embracing technologies that boost productivity. Thirteen percent of the world economy's GDP is accounted for by the building industry and related spending (MGI, 2017). Over the past 20 years, the construction industry's annual productivity growth has climbed by about 1% annually, whereas other sectors have continually boosted productivity at considerable levels (MGI, 2017). IR 4.0–related technologies have been discussed in the previous sections more extensively; therefore, this section will specifically provide the utilization of IR 4.0 emerging technologies in the context of the construction industry. These include the following:

3.2.1.1 Drone Technology

A prime example of a cutting-edge technological advancement that can considerably shorten the time it takes to complete a construction task while requiring less capital expenditure and labor is the use of drones. There are four main categories in which construction companies might benefit from drone surveying. Improvements in productivity, cost savings, higher safety regulations, and more accurate data. Drones are useful for many kinds of construction projects since they provide an airborne perspective and data acquisition. These UAVs are appropriate for a variety of tasks since they may be fitted with cameras, GPS devices, heat sensors, and infrared detectors to collect important data on the job site. In the years to come, drone technology will no doubt gain greater traction and completely revolutionize the construction business, where it has already made a significant impact.

> **EXAMPLE 3.1 THOUGHT BOX FOR STUDENTS**
>
> Without accounting for litigation, medical expenses, or reimbursement, the average cost of missed project time caused by just one accident on a construction project is USD $35,000. Taking this into account, an efficient, highly automated drone technology frees up surveyors from busy sites and quickly pays for itself.

3.2.1.2 BIM

A BIM, or Building Information Modeling, refers to a digital representation of a building or site that allows for the simulation and evaluation of any aspect of the design prior to the commencement of construction. It offers the knowledge and resources necessary for the effective planning, designing, building, and administration of structures and infrastructure. BIM adoption has risen quickly because of all the benefits it can provide. Greater access to reliable as-built data makes it easier to estimate costs accurately and reduces errors across the whole design and construction process, including shipment, maintenance, demolition, and the recycling of materials.

The BIM process and associated data formats are most thoroughly defined globally in the ISO 19650 and 12006 suites of standards. In the early stages of a BIM project, a collaborative team is assembled. It decides on the technique and data structures to ensure that the developed design information is properly organized and will be of the greatest value to those in charge of the building and operation stages. By including others who will be involved later in the project (like vendors or the client's team), this humble beginning can be greatly aided. When the project enters the execution stage, the data acquired can be used to more efficiently design and construct. The mutually accepted method can be implemented in a transparent and recorded manner when modifications to the design are required. When the building project is complete and the in-use phase starts, the data acquired from the model can then be used to operate the developed asset. A similar "digital twin" of some built asset elements is made using real-time data on the asset's functionality.

• BIM Levels

The construction industry has created and utilized numerous BIM standards and protocols throughout the last ten years. BIM can be distinguished into four levels:

Level 0: Uncontrolled computer-aided design.

Level 1: Management of 2D or 3D design.

Level 2: Data-driven 3D system using several modeling methodologies.

Level 3: An integrated, online project framework with details on procedures, finances, and life cycle management.

Given the importance of these BIM levels, several requirements must be incorporated into them when BIM is implemented on construction projects. Figure 3.1 displays these needs at various levels. Level

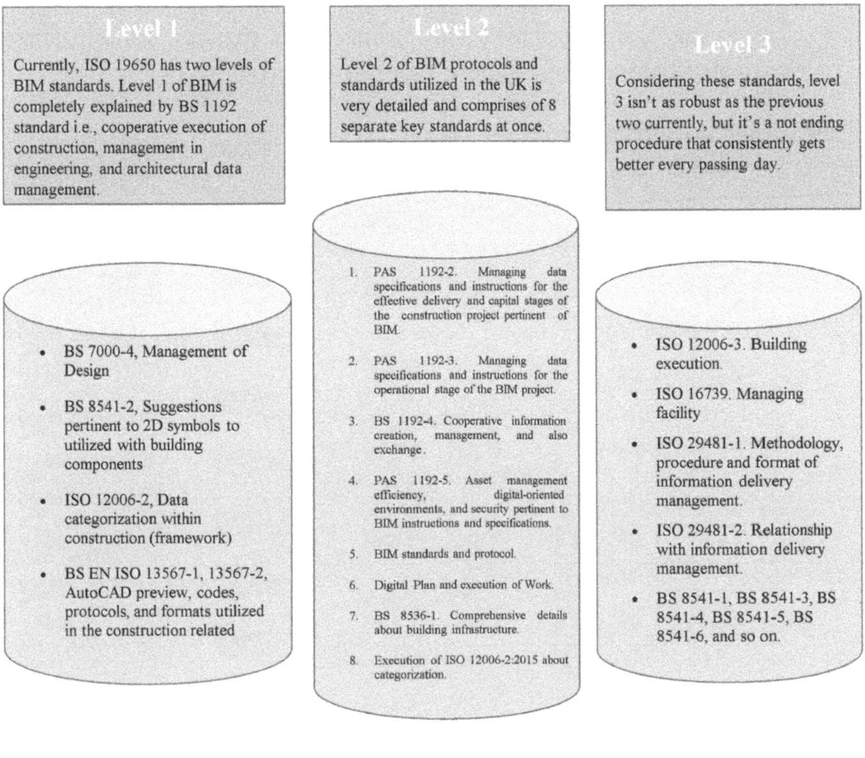

Level 1	Level 2	Level 3
Currently, ISO 19650 has two levels of BIM standards. Level 1 of BIM is completely explained by BS 1192 standard i.e., cooperative execution of construction, management in engineering, and architectural data management.	Level 2 of BIM protocols and standards utilized in the UK is very detailed and comprises of 8 separate key standards at once.	Considering these standards, level 3 isn't as robust as the previous two currently, but it's a not ending procedure that consistently gets better every passing day.

Level 2:
1. PAS 1192-2. Managing data specifications and instructions for the effective delivery and capital stages of the construction project pertinent of BIM.
2. PAS 1192-3. Managing data specifications and instructions for the operational stage of the BIM project.
3. BS 1192-4. Cooperative information creation, management, and also exchange.
4. PAS 1192-5. Asset management efficiency, digital-oriented environments, and security pertinent to BIM instructions and specifications.
5. BIM standards and protocol.
6. Digital Plan and execution of Work.
7. BS 8536-1. Comprehensive details about building infrastructure.
8. Execution of ISO 12006-2:2015 about categorization.

Level 1:
• BS 7000-4, Management of Design
• BS 8541-2, Suggestions pertinent to 2D symbols to utilized with building components
• ISO 12006-2, Data categorization within construction (framework)
• BS EN ISO 13567-1, 13567-2, AutoCAD preview, codes, protocols, and formats utilized in the construction related

Level 3:
• ISO 12006-3. Building execution.
• ISO 16739. Managing facility
• ISO 29481-1. Methodology, procedure and format of information delivery management.
• ISO 29481-2. Relationship with information delivery management.
• BS 8541-1, BS 8541-3, BS 8541-4, BS 8541-5, BS 8541-6, and so on.

FIGURE 3.1   BIM Levels.

0 is not depicted in Figure 3.1 since it is no longer a widespread practice and is rarely used in construction projects. Developed countries like the UK, USA, Australia, etc. are currently using level 2 of BIM protocols and standards. Level 3 is still in the research phase and will be available for implementation once ready for commercial use.

- BIM Dimensions

The degrees of information in a particular set of BIM data are referred to as the BIM dimensions. Figure 3.2 highlights the BIM dimensions currently available in the published literature. 6D to 10D BIM are not commonly used in construction projects due to the complexity of data management. Whereas 4D and 5D BIM are not commonly used in developing countries due to the unavailability of a skillful workforce and financial constraints. Moreover, 2D and 3D are the most used globally, whereas 1D has become obsolete and is not used nowadays in construction projects.

- Conveying the important information to project stakeholders through BIM; a project life cycle perspective

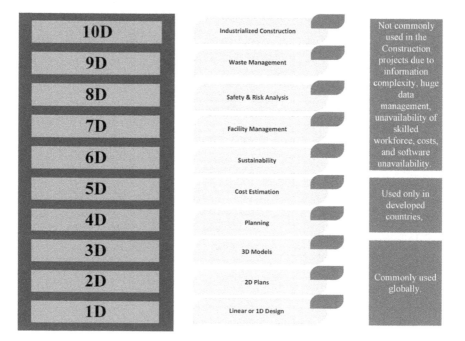

FIGURE 3.2   BIM Dimensions.

To understand and show the relationship between various types of data and information for all the engaged project stakeholders, the effective enforcement and execution of BIM protocols, guidelines, and their supporting business procedures require vigilant and thorough representation through the development of process maps or flowcharts. The suggested flowchart in Figure 3.3 has been created as a tool that works well in informing the project stakeholders concerned about the crucial information. For a smooth flow of construction activities, the generated flowchart outlines the order of the documents and the interactions that should be communicated to the stakeholders. Moreover, information dissemination should be consistently shared with the stakeholders. This will help gain the trust and confidence of the stakeholders.

The information that needs to be provided to the stakeholder has been divided according to the project life cycle phases, that is, initiation, planning, execution, monitoring and control, and handing over. Moreover, in each phase, the nature of documentation varies. For instance, when we are in the initiation phase such as strategic definition, preparation and brief, concept design, and finalized technical design, Assets Information Requirements (AIR), Employer's Information Requirements (EIR), Post-Contract BIM Execution Plan (BEP) for design team, and main contractor should be shared with the concerned stakeholders. Similarly, during the execution phase, a finalized and approved BEP should be communicated to the main contractor. Finally, when the project goes into the handing over and operational phase, BEP and AIR should be shared for verification from the client side. Furthermore, in a similar fashion, tasks and information exchange should be carried out in each phase of the project.

- Achieving integrated project management (IPD) through BIM

IPD, or integrated project delivery, is a term used in lean construction and design. IPD is a process for delivering construction projects that unites the main players in the conceptualization, fabrication, and execution stages of a project under a single contract. The goal of integrated project delivery (IPD) is to optimize the efficiency throughout all stages of design, manufacturing, and execution while maximizing project results, increasing value to the owner, and reducing waste BIM can play a crucial role in achieving the IPD

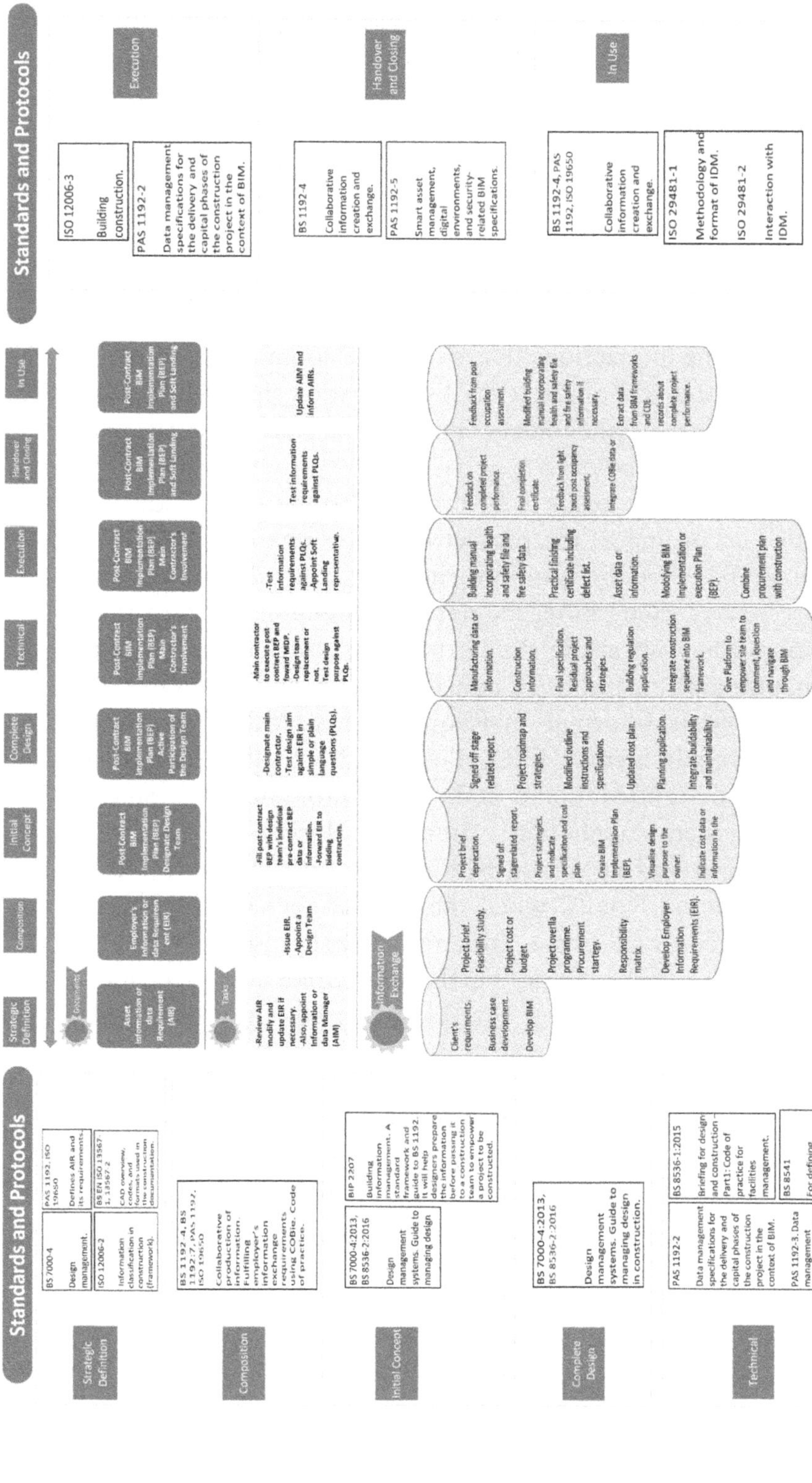

FIGURE 3.3 Information Process Chart for Engaging Project Stakeholders.

in construction projects. We have proposed a comprehensive BIM process protocol framework in Figure 3.3 that can be employed in construction projects for optimizing the project outcomes with the integration of IPD.

By ensuring that project data and information are openly accessible to all interested project stakeholders, project data and statistics can be managed effectively. Getting a high level of project homogeneity and integration is essential. This can be done via complete coordination among the clients, relevant consultants, contractors, small subcontractors, and suppliers, initiating from the early design phases. By encouraging a greater degree of coordination among all stakeholders, the use of BIM protocols and IPD will provide clients with exceptional benefits. Furthermore, it is important to carefully represent the application of BIM protocols and the business practices that support them through the development of process maps to consider and emphasize the relationships between various types of data and information for the customer and the corresponding project stakeholders.

The proposed flowchart in Figure 3.4 of process protocol (BIM oriented) has been developed as a tool that can be used to address all different and concerned project stakeholders who are interested in the procedure (from choosing an execution strategy to producing BIM documents for the project's facility management). Moreover, the widely accepted viewpoint concentrates on the concept of IPD and makes use of BIM protocols as the technological tool that will aid coordination among various project stakeholders, including the client organization and those stakeholders hired to carry out the design and construction of any facility. The proposed process protocol flowchart's important features are as follows:

- Provide a setting that is feasible and effective for the setting up and carrying out of the best strategy, provided and proposed for client organizations to ensure optimum implementation of BIM protocols/IPD.

- Map the application of BIM protocols to practices that could facilitate and hasten BIM implementation for the benefit of the client's demands and concerns.

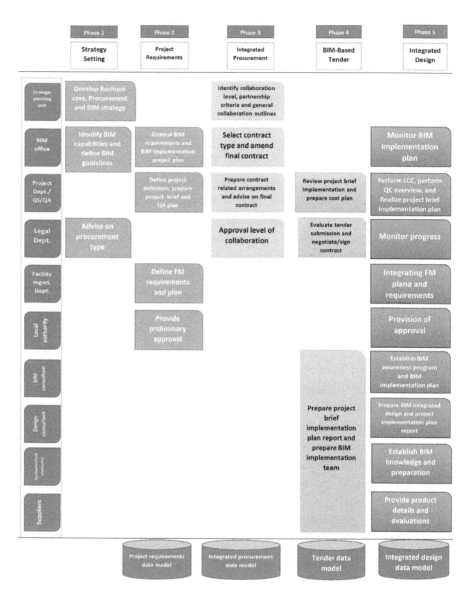

FIGURE 3.4(A) Integrated BIM Process Protocol Framework for Construction Organizations.

- Give the client organization permission to produce accurate, as-built, quality-oriented BIM models with appropriate associations to data on all structural systems and modules.

- By working together and carrying out the established design, you can increase the delivery's progress through consistency and efficiency, both in the design and implementation stages.

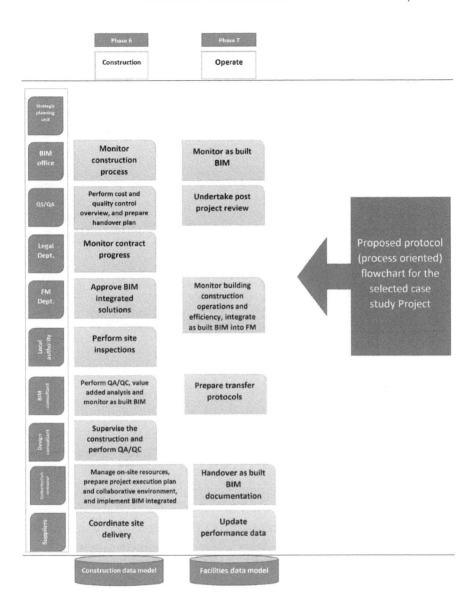

FIGURE 3.4(B) Integrated BIM Process Protocol Framework for Construction Organizations.

3.2.1.3 Prefabrication and Modularization

Building components are manufactured off-site in controlled conditions using prefabrication and modularization construction techniques and then transported to the construction location for installation. Modular construction typically involves creating standardized building components in a factory off-site before assembling them on-site. The words "prefabrication," "modular construction," and "off-site construction"

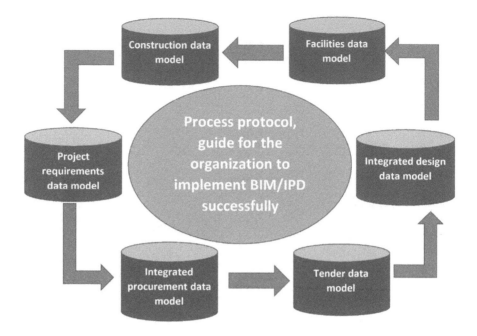

FIGURE 3.4(C)   Integrated BIM Process Protocol Framework for Construction Organizations.

are often used interchangeably. These terms cover a wide range of techniques and configurations, from straightforward parts that are joined and interfaced to more complex 3-D volumetric installations with full accessories. Numerous factors imply that modular building might have unheard-of lasting significance. But in contrast, there has recently been a surge in investment and interest in it. Digital technology advancements, which have, among other things, streamlined module design and enhanced delivery logistics, have profoundly changed the modular building approach. Prefab home attitudes are beginning to change, especially as new, more intriguing material alternatives improve the aesthetic appeal of the buildings. Perhaps most importantly, the CEOs of the construction sector are adapting their thinking. Many executives are becoming aware that it could be time to change their roles as competitors with a technological edge to enter the market.

Moreover, only a handful of places, such as Scandinavia and Japan, have successfully established prefabricated houses. It has fluctuated in popularity during the postwar era in markets, including the United Kingdom and the United States. There is evidence to assume the current rebound may be distinctive, but modular building in the

European and US markets can provide yearly savings of up to $22 billion. The industry is embracing new, lightweight materials as well as digital innovations that increase design flexibility and capability, raise manufacturing accuracy and productivity, and simplify logistics.

3.2.1.4 Green Construction

The term "green construction" refers to all the procedures used to create and utilize the built environments in a way that is as non-harmful and environmentally friendly as feasible. Green building is concerned with lessening the bad effects on the environment while enhancing some good effects, from the initial design phases to assembly, the functionality of the project after completion, and later removal. Due to the carbon-intensive nature of the construction business, green techniques and solutions are essential to lowering the CO_2 emissions of the construction industry.

A conceptual diagram of the life cycle of a green and sustainable building is shown in Figure 3.5, along with the key elements of sustainable building (such as electricity, water, debris, and materials) associated with each stage. Green buildings try to reduce the adverse impacts and hazards to the environment during each stage of their life cycle by considering and evaluating the environmental consequences of each step. Since the stages are linked to several stakeholders (such as building designers, builders, end users, etc.) with various specialties and objectives, it is essential to guarantee that everyone is working together to uphold environmental and sustainability standards.

3.2.1.5 Advanced Materials

Advanced materials are those that have been produced specifically to have better or novel qualities that offer greater efficiency in comparison to standard materials. The emergence of new types of materials is a result of the quick advancement of technology. Customers seek out building materials with high engineering qualities, which are abundant. After substantial investigation, advanced building materials were created. They raise the level of strength and increase durability. Thermal cracking ought to be avoided. Engineers and other construction specialists are constantly searching for better materials to increase durability. Furthermore, increasing strength is also a simple process. Following are some instances of advanced materials that can be utilized in construction projects:

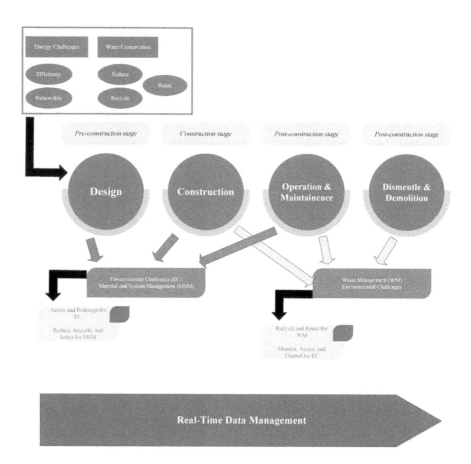

FIGURE 3.5   Green Construction Attainment Framework.

- Light-emitting concrete
- Self-healing materials
- Cross-laminated timber
- Bioplastic
- Carbon fiber
- Translucent wood
- Transparent aluminum
- Fiber-reinforced concrete
- Porous concrete

- Self-compacting materials

- Ultra-high-performance concrete

- Polyethylene terephthalate for soil stabilization

- 3D printed graphene

3.2.1.6 AVR

Architects may elevate their ideas to an entirely new level with the aid of augmented and virtual reality (AR/VR), opening a wealth of opportunities for the construction industry. Effectively integrating new technologies gives construction companies a competitive advantage. Because of this, AR/VR technology can advance inventors. By integrating BIM technology and immersive VR headsets, architects and designers can gain a better knowledge of an area before it is really built and thus enhance their strategies for developing it. By giving stakeholders a virtual tour of the future area rather than a 2D depiction or model, companies may now provide a stronger case when competing for a new project. Other key uses include the following:

- Visualization of planning

- Monitoring and controlling real-time project data

- Safety management

- Better communication and information circulation between project teams

- Remote maintenance

EXAMPLE 3.2 THOUGHT BOX FOR BUSINESSES

Global building investment is expected to reach $15.5 trillion by 2030. By 2028, it is expected that AR/VR technology would hold a $250 billion market share, which might lead to a complete industry revolution.

On the other hand, the most recent AR/VR technology can cut building expenses by up to 90% and save the construction industry up to $15.8 billion by preventing accidents caused by inaccurate or missing data.

Source: (P.P., 2022)

CASE STUDY 1

Smart Helmet-DAQRI

The DAQRI Smart headgear is a head protection that allows users to view projects and 3D renderings in augmented reality as a vast and immersive 3D environment. Individuals can increase work efficiency by comparing their current work to the original idea.

CASE STUDY 2

Dalux

Dalux offers three augmented reality options for civil construction.

Visitors can create a 3D model and electronic drawings on the real construction site using the free iOS and Android software Dalux Viewer.

Dalux Build uses augmented reality in conjunction with additional resources to streamline the management of construction projects and to provide a steady stream of data.

On the other hand, DaluxFM offers facility management strategies, including planning and management of assets, as well as contracts for delegated and related services, electrical servicing, hydraulics, rental management, and site conservation tasks.

CASE STUDY 3

Fologram

Fologram converts 3D models into scale-model construction guidelines using augmented reality goggles like the HoloLens. Through digital directions that are virtually imposed on the workspace and lead masons step-by-step during the construction phase, the program seeks to ease the construction of challenging projects that involve several measurements, verifications, along with particular care.

3.2.1.7 AI

AI in the construction industry has the potential to significantly alter the sector and foster widespread innovation and creativity. Without overstatement, we can assert that this cutting-edge technology ensures rapid and precise outcomes. Of course, it's yet unclear how artificial intelligence will develop in this field. However, AI is now an essential and cutting-edge technology for any firm due to its distinctive properties and uses in the construction industry. AI is a field of study that gives machines the ability to reason, acquire knowledge, and act on their own. It can be utilized to streamline building projects and procedures by automating tasks and providing insights. AI may also be employed to detect possible safety risks before they materialize, increasing workplace security. For instance, AI can be used to locate workers who aren't using safety equipment on a building site. It may notify the site manager to take appropriate action.

Project management is more difficult because of the shortage of digital competence and the operational character of the construction sector. This is why we encounter issues with cost ineffectiveness, project delivery complications, inadequate execution, faulty decision-making, and deficiencies in productivity, health, and safety systems in most architectural projects. These issues in construction projects can be greatly resolved by using artificial intelligence.

3.2.1.8 3D Printing

The automated method of producing individual construction components or full structures using a 3D printer is known as 3D construction printing (3DCP or 3DP). To be more precise, 3DCP is also known as additive manufacturing or additive construction since building components are printed layer by layer, as opposed to the ink that standard printers employ. Both on-site and off-site work can be done. There have been many queries about 3D printing technology since the 1980s. Nevertheless, the method's progress, which allows for the creation of three-dimensional objects by stacking consecutive layers of material, has brought it more notoriety. Buildings or living spaces, work environments, bridges, walls, modular building components, reinforcement molds, columns, urban furnishings, and even decorative elements can all be constructed using this method of construction, which is very adaptable and may be employed to construct both specific project components as well as various kinds of intricate structures in their entirety.

EXAMPLE 3.3 THOUGHT BOX FOR BUSINESSES

One of the most significant technological developments and innovations of the twenty-first century is 3D printing. In terms of market value, it was about $190 million (USD) in 2021, and by 2030, it is anticipated to be worth $680 million (USD). The environmental challenges about the waste produced by the industry, the potential for customization and architectural flexibility utilizing this technology, and the speedy production of models and prototypes are the reasons that have the biggest impact on this growth.

The capacity for embracing and expansion increases in tandem with the advancement of technology; by the year 2028, it is projected that its compound annual growth rate (CAGR) will surpass that of 2021 by 91.5%.

Source: (Iribar, n.d.)

CASE STUDIES AROUND THE WORLD IN WHICH 3D PRINTERS WERE USED IN THE CONSTRUCTION PROJECTS:

1. Apis Cor 3D-printed house in Russia
2. TECLA house construction in Italy
3. VULCAN Pavilion in UAE
4. Shanghai Chenshan Botanical Garden in China
5. The BOD in USA

3.2.1.9 Robotics

Construction is a labor-intensive industry. The construction sector is significantly behind in adopting robotics, automation, and technical improvements because it is one of the least mechanized industries. Robotics and artificial intelligence are two of the newest advancements in the building construction industry that have lately gained attention. Such technology allows for more precise and timely building while also saving time, money, and other resources. Construction experts may now reduce human error and ensure superior outcomes in a quick-moving construction process by using robotics technology.

Construction robots come in a variety of designs that are now on the market. Each of them has a unique purpose and is employed in various ways during a construction job. These include the following:

- Bricklaying robots
- Plastering robots
- Robots in 3D printing
- Assessment of construction sites through robots
- Self-driving construction vehicles
- Drones

3.2.1.10 IoT

IoT is revolutionizing the construction sector by bringing new methods of job site management, enhancing safety, and boosting productivity. The Internet of Things (IoT) is a network of connected gadgets that communicate and collect data online. These electronic devices have sensors and software built right into them, allowing them to talk to other systems and one another. Many different industries, including transportation, agriculture, and the health care sector, can benefit from the adoption of IoT devices. IoT is utilized in the construction sector to monitor job sites and machinery, keep an eye on safety, and enhance project management.

The usage of sensors is one of the most significant means that IoT is being implemented in construction. Sensors are utilized to keep track of several variables, including pressure, humidity, and temperature. Construction managers can spot possible concerns before they turn into more serious problems by keeping an eye on these parameters. To ensure that concrete cures effectively and does not fail or fracture, sensors can be employed to track its temperature and humidity as it hardens. The structural integrity of structures can also be monitored by sensors to make sure they are reliable and secure.

3.2.1.11 5G Technology

The emergence of 5G technology has completely changed how companies conduct business across all sectors, especially the construction industry. The potential of 5G technology to offer quicker and more effective communication is one of its main advantages. Construction crews can connect in real time and share massive volumes of data with ease thanks to 5G. Higher efficiency, improved teamwork, and quicker completion times are the results of this. Other benefits include the following:

- More rapid and effective communication.

- Increased safety precautions.

- More effective project management.

- The usage of automation has grown.

- Improved assessment of data.

3.2.1.12 6G Technology

The integration of 6G with several previously unconnected technologies will continue the trend started by technologies enabling 5G functions. AI, big data analytics, and next-generation computing are just a few of the crucial technologies that will combine with 6G. The performance of current 5G technologies will be improved by 6G networks, which will also broaden their support for constantly new and inventive applications in the areas of transmission, sensing, wireless cognition, and imagery.

While 5G uses microwave frequencies like mmWave, 6G will use even shorter wavelengths in the terahertz (THz) band, which operates between 100 GHz and 3 THz. While 5G has a significant influence on the Radio Access Network (RAN), 6G networks, which are primarily driven by an enormous rise in frequency, will allow the demand for antennas nearly everywhere and will have a far greater effect.

3.2.1.13 Cyber-Physical Production Systems (CPPS)

Cyber-Physical Production Systems (CPPS) are a hybrid of digital and physical technologies that improve production procedures. The usage of CPPS in the construction industry can enhance teamwork and coordination, keep an eye on tools and resources, and speed up workflows. Cyber-physical production systems merge physical and digital systems into one infrastructure. There are feedback loops where physical operations impact the computations and vice versa. Physical functions are tracked and administered by integrated computers and networks. CPPS blends the dynamics of physical processes with those of software and networking by employing metaphors and modeling, design, and analytical methodologies.

3.2.1.14 Cyber-Physical System (CPS)

Construction companies now have more opportunities to use technology to streamline construction procedures due to the development of cyber-physical systems (CPS). This is because organizations using BIM/VDC are not fully utilizing their potential. For instance, there is still a lack of synchronization and coherence between the actual construction work being done on the site and the virtual models created by designers. Virtual models are created during the design phase, but they are rarely, if ever, utilized after that in construction.

CPS offers a suitable technique for construction team members to bridge the discrepancy between virtual models and the actual construction by allowing for a close connection between computer models and their associated physical elements. Leveraging this bidirectional collaboration will make it possible to create and implement a wide range of apps, services, and other advances. It would be especially much simpler to use the virtual model as a crucial step in the actual construction process.

3.2.1.15 Digital Twin

This technology builds digital representations of actual items using AI and other self-learning algorithms, assisting professionals in gathering and analyzing data-driven analytics. The development of digital twin technology is nothing short of a miracle for the building sector. Large-scale real estate and infrastructure ventures are now frequently developed and managed using it in several nations. The technique of 3D modeling in buildings has been elevated to a whole new level by the development of digital twin technology. It gives precise information about the characteristics and conditions of physical items. Digital twins and their associated real-world objects can be kept in sync extremely easily with the use of AI and other automated technologies.

The procedure uses cloud data, survey data, and other IoT sensors that let devices transmit data back and forth. The data gathered from these many resources is then processed and combined into a 3D model, which greatly facilitates designing and construction work on a specific project. Building information models (BIMs) are a more sophisticated version of construction plans. To increase a BIM's dependability and effectiveness, data from a digital twin is supplied to it. Digital twins constitute a source of essential and analytical data when a large-scale development activity has begun, keeping the BIM planned and real processes in a desirable state of synchronization.

3.2.1.16 Cloud Computing

Project management, archived data, and collaboration are just a few of the many uses of cloud computing in the construction sector. Cloud computing has expedited project workflows and improved team collaboration by offering a centralized platform for managing data. Additionally, it has made it easier to deploy cutting-edge technology like building information modeling (BIM), which has completely changed how construction projects are planned and handled.

The importance of cloud computing in the construction industry resides in its capacity to improve data management and interaction.

It empowers teams to collaborate more effectively, minimizing delays and errors by offering an effective and adaptable infrastructure for data storage. Real-time communication is supported by cloud computing as well, allowing project managers to make decisions based on current information.

3.2.1.17 Autonomous Construction Machinery

The significance of self-driving technology is growing beyond the world of autonomous vehicles as it continues to advance. In preparation for a time when machines assume the wheel, the construction sector is now investigating the incorporation of autonomous technology into construction vehicles and equipment. In recent years, self-driving technology has advanced significantly, and its use in construction equipment is already a possibility. Construction vehicles and equipment can travel and function independently thanks to cutting-edge sensors, machine-learning algorithms, and real-time data processing. Without human assistance, these machines can comprehend the environment, make deft decisions, and carry out tasks.

EXAMPLE 3.4 THOUGHT BOX FOR BUSINESSES

Autonomous construction equipment market is expected to grow from $11.861 billion in 2020 to $31.841 billion in 2030.

3.2.1.18 Blockchain

With the help of blockchain technology, data may be recorded in a way that makes it difficult or nearly impossible to change, hack, or cheat. By streamlining the procurement procedure, blockchain might lessen the high degree of fragmentation and complication that characterizes large-scale projects. The materials' provenance can decrease waste and advance the quality of goods and services with high transparency. Such technologies can improve reliability, both in terms of procurement and overall project delivery. Blockchain has the potential to become the single source of information for all elements of a construction project when combined with BIM (building information modeling). A trusted digital twin of an asset can be created using such a model, facilitating not only its design and construction but also its maintenance and functioning during the asset's entire existence.

The potential to transform the construction sector for the better has become too big to ignore, even though the technology is new and there are still several early challenges to overcome. Following are some of the challenges of implementing blockchain technology in the construction industry:

- Rules and regulations for organizations.

- Standards and protocols to implement blockchain technology.

- Data privacy.

- Data management.

- Shortage of skilled workforce in the context of blockchain technology.

- Fear of change.

- Stakeholder's trust.

- Operational challenges.

- Financial barriers.

- Social challenges.

- Difficult user interface.

3.2.2 Incorporation of IR 4.0 in Construction

The construction industry has also undergone a profound transition and made significant progress in embracing technology improvements and transforming the status quo. Construction researchers started looking into the possibility of incorporating Industry 4.0 into construction after being influenced by the advantages of the fourth industrial revolution. Despite many comparisons between manufacturing and building, the former is pursuing and embracing Industry 4.0 from a different standpoint. The demand for construction 4.0 is being fueled by the quick rise of novel technologies, the significant change in owner expectations, the movement to customized products, and the requirement for sustainable and green structures. Figure 3.6 shows a four-layer execution plan for the application of Construction 4.0 that is based on the concept of IR 4.0. Construction 4.0 and nine of its most frequently referenced technologies are introduced in the first layer, illustrated by Figure 3.6. The second layer describes how Construction 4.0 technology will be included in the lifecycle of a construction project and the creation of a plan for integrating Construction 4.0 technology across the project life cycle. The third layer investigates how Construction 4.0 technologies are connected and integrated. The prerequisites necessary to achieve Construction 4.0 are included in the fourth tier of the execution plan.

3.2.3 Intelligent Construction Hurdles

3.2.3.1 Project Ingenuity

Since each construction project is different, therefore the significance of creativity and capability to think innovatively and present out of the box solutions to solve complex and dynamic project problems has increased manifold. Typically, in other industries, you would develop the same product and instruct a machine or piece of technology to create an enormous quantity of the same product, but in the construction industry, every site is unique, and every building is made for a particular purpose, a particular environment, and a particular use. Even though this is difficult given the state of technology, it is best if every component is modular and reusable when creating an integrated intelligent system.

3.2.3.2 Diversity

Construction operations take time to complete. They are intricate and include numerous stakeholders, contractors, and subcontractors. The interconnectedness of all operations and the frequent sharing of resources and labor force all contribute to the complexity of construction projects. Most of the time, the information does not move sensibly and precisely down to the lower hierarchies.

3.2.3.3 Risks and Unpredictability

In the context of the fourth industrial revolution, a concept known as the "smart factory" is developed, where cloud computing and cognitive computing gather data and determine actions to boost productivity. BIM, which was developed from computer-aided design (CAD), is a cutting-edge technology that assists a construction project throughout its entire life cycle by offering a virtual model and pertinent building data. It facilitates the visualization of demands and requirements, thus lowering the risk factor.

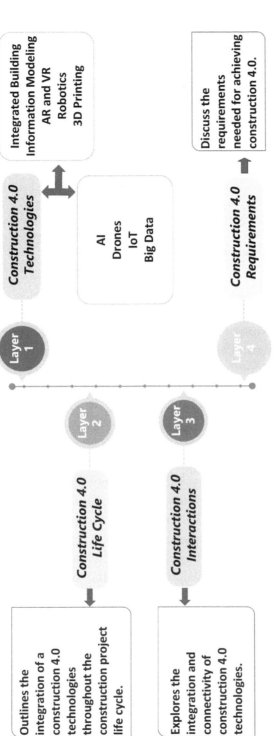

FIGURE 3.6 Implementation Plan of the Industrial Revolution in the Construction Industry.

EXAMPLE 3.5 THOUGHT BOX FOR STUDENTS

Students have many opportunities to contribute to and influence the direction of the construction industry through the utilization of Industry 4.0, known as Construction 4.0. The following are some strategies students can use to help the construction industry adopt Industry 4.0 technologies and practices:

1. Develop your understanding of digital construction technologies.
2. Work together on research projects.
3. Take part in internships and programs that are related to the industry.
4. Donate to organizations and industry associations.
5. Examine the concepts of entrepreneurship and innovation.
6. Encourage the use of smart, sustainable construction methods.
7. Take advantage of lifelong learning and stay current.

3.3 CONSTRUCTION 4.0 IMPLEMENTATION DRIVERS

The implementation of new technology significantly depends on some key driving factors. Therefore, the subsequent section will highlight the drives in the context of Construction 4.0 implementation.

3.3.1 Construction 4.0 and Project Life Cycle

The utilization of Construction 4.0 technologies is viewed from a life cycle perspective at the first level of the integration activities. After the initial planning stage, a project proceeds to the design, execution, and facilities management phases. When a technology is implemented where appropriate over the whole life cycle of a construction project, its full prospects are realized. In the construction sector, interest in the BIM and robotics potential synergies has grown. The advantages of utilizing 3D printing with robotics to automate different building jobs include lower safety risks, better schedule control, and lower construction costs. This is because robotics and 3D printing have distinct applications in the construction sector. When managing construction projects in the past, construction organizations frequently used a project-thinking approach.

3.3.2 Mindset Transition and Training

The transition to Construction 4.0 necessitates a change in methods and ways of thinking. The evolution of digital technology and its customization to the requirements of the building industry are both extremely dynamic and pricey processes. Employees and workers will also require additional training when new technologies are deployed. As a result, only if the new technologies can be employed on many projects and are incorporated into the company's processes can construction companies provide surplus value. In contrast to the conventional project-oriented strategy, the shift to Construction 4.0 requires a mindset that is process-oriented. This attitude shift, however, forces construction organizations to digitize their current processes. This poses an additional challenge because 1) most systems and processes were created before modern digital tools were available, and 2) not all processes can be directly automatized. As a result, all current processes must be reengineered to account for the change in perspective and enable the IR 4.0 of the construction industry.

3.3.3 Strategy Development by Organizations

To successfully transition to Construction 4.0, construction firms need a strategy for the digitization change management process that meets the goals of the organization. This is crucial since Construction 4.0 encompasses more than just technology. It involves altering the perspectives of everyone involved, from management to field staff, as well as the procedures under evaluation. Thus, having a clear vision that these people share is crucial. Companies can only switch from traditional project thinking to process thinking after these conditions are satisfied. The absence of international standards and a framework for implementation, in addition to the difficulties involved in transitioning construction towards a process-thinking business, is another barrier. Additionally, because Building 4.0 depends so heavily on various information technology (IT) technologies, data security and cybersecurity concerns must be resolved, and stringent security standards must be implemented. To distribute risks among stakeholders, it is also necessary to talk about legal and contractual issues.

Organizations should develop their strategic goals under three levels, that is, strategic level, tactical level, and operational level, as shown in Figure 3.7. The strategic level focuses on the long-term planning of activities that should be aligned with the company's vision and missions. Similarly, the tactical level focuses on short-term level executive management of operations and provides policy directions for attaining the strategic level subgoals. Finally, the operational level deals with micro-level details and plans for day-to-day activities that assist in smoothly running the business activities.

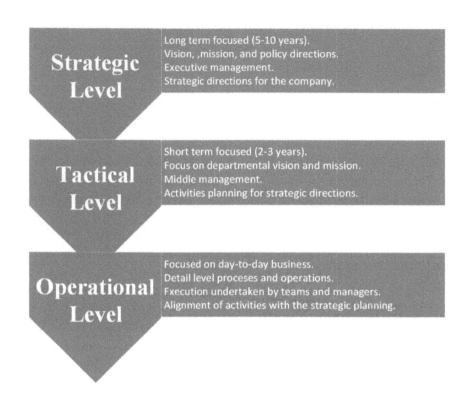

FIGURE 3.7 Strategy Development for Organizations.

3.3.4 International Consensus on Construction 4.0

Moreover, international agreement on what Construction 4.0 entails is still lacking. It is seen as a means of identifying coordinated synergies between the key changing technology approaches in the sector, whereas it can also be described as a pure and sensible application of Industry 4.0 in the construction sector (i.e., utilization of widely available connectivity technologies for real-time decision-making). However, it is a more all-encompassing approach that goes beyond the fundamental technology foundation to better solve the industry's current challenges. Whatever the definition, the main change brought about by Construction 4.0 seems to focus on a decentralized connection between the real world and the digital world through continuous interconnection. The bridge between these two worlds in the building industry already exists because of technological solutions like BIM. It is possible, for instance, to use thorough modeling methods for a construction project and even to create two-way communication between the models and the construction site. But a person needs to be there to manage and maintain this communication. As Construction 4.0 takes hold, a variety of technologies will eventually take over this function played by humans to eliminate human intervention and proceed toward a decentralized synthesis of physical reality and its representation in cyberspace.

3.3.5 Digitized Designs

It is feasible to provide individualized, intelligent, and linked building products in the Construction 4.0 age. Through digitalized design, building, and operation, the industry is being transformed in this way. The construction industry is undergoing a paradigm shift because of this digital transition. This paradigm shift, which has been extensively studied in the scientific literature concerning the diffusion of BIM, must occur on a variety of dimensions, including those of technology, organization, policies, etc. The issue of data and its management will be crucial in the context of Construction 4.0 and force the industry to adopt new business models. A sizable amount will be produced through sensors and CPS systems, in addition to the vast array of data and information that is typically generated during the development and execution of construction projects. The interaction between Industry 4.0's benefits and drawbacks for the construction industry and BIM's function as the planning domain determines the fundamental structure of the CPS (Maskuriy et al., 2019). Having stated that, Figure 3.8 illustrates how the planning domain governs the physical and digital domains. When physical domains and cyber domains work in a synchronized way, the BIM can provide optimized and highly efficient outputs because the data obtained from both domains will assist in enhancing the work environment for both the operator and the software.

3.4 CHALLENGES OF APPLICATION OF INDUSTRIAL REVOLUTION IN THE CONSTRUCTION INDUSTRY

The technical advancements are taking place at a rapid speed, which should motivate the construction stakeholders and top management to align their vision, mission, goals, and strategic planning by complying with the requirements of the shifting technologies. In this context, IR has brought revolutionary change and opportunities for the construction industry. But these opportunities also bring certain

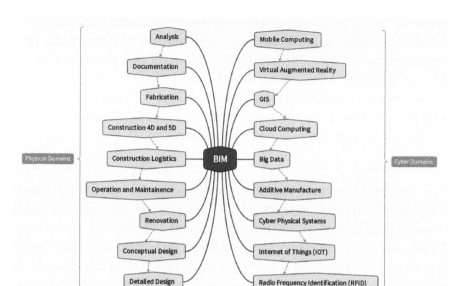

FIGURE 3.8   Interrelation of Physical and Cyber Domains in Construction 4.0.

EXAMPLE 3.6 THOUGHT BOX FOR STAKEHOLDERS

Construction 4.0 offers several advantages, but those who are involved in construction projects should be aware of any risks and challenges that can present themselves. Here are some concerns to be aware of:

1. Costs of application.
2. Intricacy in technology.
3. Data protection and confidentiality.
4. Workforce transformation.
5. Aversion to change.
6. Regulation adherence.
7. Collaboration across disciplines.
8. Expansion and sustainability over the long term.

challenges with them that can obstruct the implementation of those technologies. The subsequent section will discuss this aspect in detail.

3.4.1 Problem of Cybersecurity

The way we live and work is changing because of the fourth industrial revolution, commonly described as Industry 4.0. It is distinguished by the incorporation of cutting-edge technology including robotics, IoT, AI, and big data analytics. Industry 4.0 has the potential to bring about a wide range of advantages, but to fully fulfill this potential, several issues must be resolved. The problem of cybersecurity is one of Industry 4.0's main difficulties. Cyberattacks are more frequent and sophisticated as connected devices and data are used more frequently. Both individuals and businesses are susceptible to monetary loss, data breaches, and reputational harm. Businesses need to be proactive in securing their systems and data, which includes making investments in reliable software and IT infrastructure and putting in place thorough security policies. The implementation of Industry 4.0 offers both enterprises and individuals several exciting potentials, but many problems must be overcome. These include interoperability and standardization, employment displacement, cybersecurity, and ethical and societal repercussions. Governments, companies, and individuals may maximize the benefits of Industry 4.0 while lowering the dangers and difficulties involved by cooperating. Figure 3.9 illustrates the key elements of a cybersecurity environment. Whenever a malware attack happens, inbuilt cybersecurity features such as firewalls play a crucial role in saving data and monetary losses.

3.4.2 Adoption and Maintenance of IR 4.0 Technologies

The adoption and maintenance of IR 4.0 technologies in the construction sector are thought to be excessively expensive, making innovation less likely. The experts in the field oppose changing their conventional practices and show no interest in embracing new technologies. The reasons for this lack of interest in implementing IR 4.0 approaches are the lack of technical know-how and

FIGURE 3.9 Cybersecurity Elements.

specialized staff and the client's unwillingness to insist on using the technology.

3.4.3 Hesitation of Organizations and Inadequate Supportive Facilities

Inadequate electricity, financial constraints, difficult connection to wireless broadband, and inadequate knowledge are some of the major obstacles the construction industry must overcome. In addition, construction firms prefer to stay with tried-and-true methods since they regard adopting cutting-edge technology as a risk. Most firms do not use IR technologies because of the small to medium-sized projects they work on. Several businesses are debating whether to embrace technology assets due to the cost of integrating and sustaining them. Yet the industry lacks the adaptability to swiftly adopt new technologies.

3.4.4 Shortage of Skilled Workforce

Studies show that the labor market is unbalanced in terms of worker supply and demand, and there is a lack of education in the sector. In addition to the shortage of digital skills, many firms worry that digitization could lead to job losses. Implementing IR 4.0 is challenging because of concerns over job loss.

3.4.5 Insufficient Technical Competence and the Unavailability of Laws and Regulations

Insufficient technical competence and the unavailability of laws and regulations governing the deployment of IR 4.0 technologies pose another barrier to adoption. Although the civil construction industry should move to the digital world, doing so will be challenging for e-governance. The constraints include creating cheap access policies, creating a wireless broadband infrastructure, and having the necessary capabilities to create e-government services. The high expense of gaining innovation and the high training expenses are also major obstacles to deploying the technology.

3.4.6 Accountability, Transparency, and Privacy Issues

Artificial intelligence and other cutting-edge technology pose concerns about accountability, transparency, and privacy. The algorithms and data sets that are utilized to make judgments have the

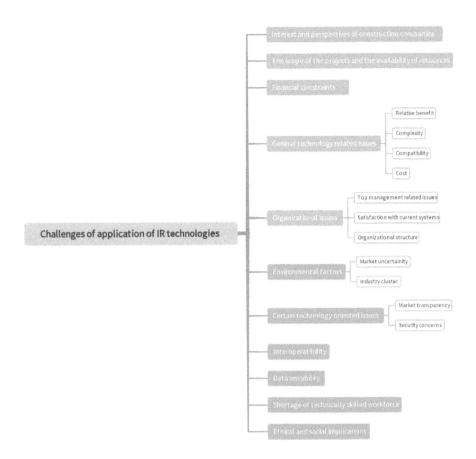

FIGURE 3.10 Challenges in the Implementation of IR Technologies in Construction.

potential to be biased and discriminatory. Companies must consider their social duty as well as the long-term effects of their actions on the community and the environment. It can be difficult for businesses to effectively integrate and run these systems given the variety of technologies and platforms being employed.

3.4.7 Lack of Interoperability and Standards

The potential for innovation, cooperation, and value creation across industries is also constrained by a lack of interoperability and standards. Moreover, Figure 3.10 summarizes the key challenges in the application of IR technologies in the construction industry.

REFERENCES

Iribar, I. n.d. *3D Printing in Construction. How Does It Work?* [Online]. CEMEX Ventures. Available: https://www.cemexventures.com/3d-printing-in-construction/ [Accessed 2 August 2023].

Maskuriy, R., Selamat, A., Ali, K. N., Maresova, P. & Krejcar, O. 2019. Industry 4.0 for the construction industry—how ready is the industry? *Applied Sciences*, 9, 2819.

MGI. 2017. *Reinventing Construction: A Route to Higher Productivity*. McKinsey Global Institute. https://www.modular.org/2019/07/20/reinventing-construction-a-route-to-higher-productivity/#:~:text=The%20global%20construction%20industry%20has,of%20the%20rate%20in%20manufacturing

P.P. 2022. *How Can AR/VR Benefit the Construction Industry?* [Online]. PixelPlex. Available: https://pixelplex.io/blog/ar-vr-in-construction/ [Accessed 2 August 2023].

CHAPTER 4

Project Management Knowledge Areas

4.1 PROJECT MANAGEMENT PHILOSOPHY IMPLEMENTATION IN CONSTRUCTION

Project management is the adoption of techniques, methods, processes, and skills within the context of a project to successfully achieve its desired predefined goals and objectives from its initiation till its closure phase. It not only includes multiple tools and techniques but also diverse teams, who work together in the projects to attain the end goal of the project defined by clients at the start. Therefore, the significance of project management in the construction industry cannot be overlooked as every project is unique with native risks and unforeseen circumstances. Hence, this uniqueness of construction projects makes the implementation of project management philosophy mandatory as projects have evolved to become more complex. This complexity demands the utilization of advanced techniques and methods.

Moreover, projects are becoming more complicated, and as a result, project management knowledge is also evolving at a rapid pace. Currently, according to the Project Management Body of Knowledge (PMBOK), there are 13 project management knowledge areas with 49 processes (PMI, 2017) as presented in Figure 4.1, and without the integration of all these in the projects, it becomes very hard to

DOI: 10.1201/9781032621760-4

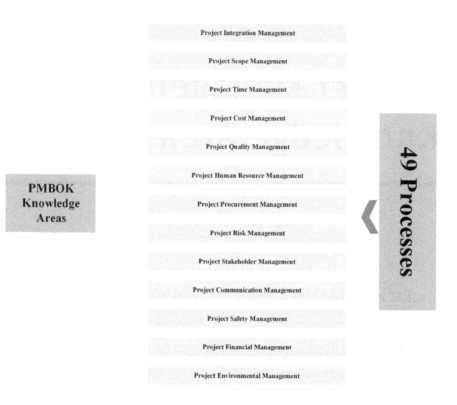

FIGURE 4.1   Project Management Knowledge Areas.

accomplish the goals of the project within the specified time, cost, resources, and quality.

Similarly, these project management knowledge areas should be implemented efficiently throughout all the life cycle phases of any project. A construction project, irrespective of its nature, passes through five life cycle phases, that is, initiation, planning, execution, or implementation, monitoring and control, and closure, as shown in Figure 4.2. A project's success can be defined based on the outcomes from these life cycle phases, and the outcomes greatly rely on the utilization of tools, techniques, methods, and procedures.

It can be seen in Figure 4.2 that all the processes in the life cycle phases are interconnected with each other. The final product of the initiation phase becomes the starting point of the planning phase, the outcome of the planning phase becomes the starting point of the execution phase, and the output of the execution phase becomes the input of the closure phase. Moreover, monitoring and control

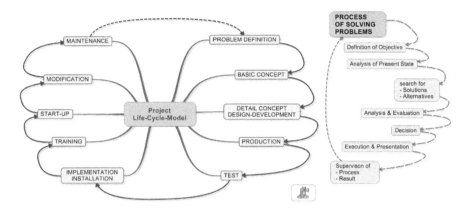

FIGURE 4.2   Project Life Cycles in Project Management.

phases remain active in all the phases of a project. Furthermore, the process of problem-solving and life cycle phases go hand in hand. Construction projects are unique, with a diverse mix of involved stakeholders with varying interests and demands. The probability of occurrence of a problem is high, and its resolution is mandatory for the successful execution of a project in all the life cycle phases.

4.2 IR 4.0 AS A SOLUTION FOR PROJECT MANAGEMENT KNOWLEDGE AREAS

The key aim of project management is to organize and manage the available resources with a focus on the client's needs and expectations and to complete the projects within the constraints of time, cost, quality, safety, and environmental protection. Also, the outcome of any project is to produce a unique product, service, or result. Owing to this characteristic of construction projects, a complete transition in all the life cycle phases of a project needs due attention from all the key stakeholders around the world. The reason lies in the fact that development in automation and technology is moving to higher levels with each passing day. To cope with this technological revolution, the construction industry players need to align their efforts for the successful attainment and implementation of construction projects.

4.2.1 Traditional Project Management Skills

Traditional project management practices have been utilized in the construction industry for decades for managing big and complex projects. Resources from diverse skill areas are gathered to attain the common end goals of the projects. Project targets are very carefully drawn, and project accomplishment criteria, that is, time, cost, and quality, are finalized. Furthermore, through project management tools and network methods, Gantt charts, critical path method, program evaluation and review technique (PERT) method, project compression techniques, and project performance measurement techniques like earned value analysis (EVM) and work breakdown structures (WBS). In traditional project management, as shown in Figure 4.3, the key competencies of the project managers can be

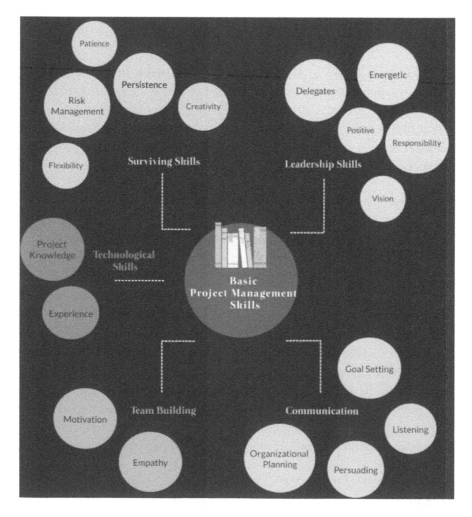

FIGURE 4.3   Project Manager's Skills in Traditional Project Management.

divided into technology, leadership, survival, communication, and team-building skills. These skills are expected from project managers in traditional project management philosophy.

4.2.2 Industry 4.0 Project Management Skills

Subsequently, it can be said that project management is an art, and project managers are artists. With larger and more complex features of construction projects, the utilization of advanced automation, simulation, and technology-oriented tools and techniques should be used. This advancement in automation and technology can be traced to the Industry 4.0 concept. Because of Industry 4.0, businesses around the globe have incorporated digital communication technologies and systems in their activities and processes. The Internet is used practically everywhere today and is largely recognized as the most important instrument for communication and information. Businesses all over the world are getting more and more digitalized because of the creation and use of virtual representations of the actual world and the increasing prevalence of cyber-physical technologies. Due to this, the way organizations are run today has undergone a substantial shift that is mostly characterized by automation, Robotics, Cloud Technology, and IoT. Moreover, IR 4.0 has proved to be providing economic, technological, and managerial benefits to businesses globally.

For fully added systems and procedures, Industry 4.0 proposes a new organizational and control pathway. Meeting unique client needs at mass manufacturing prices is the major objective. As a result, order management has an impact on all areas, including manufacturing, leasing, distribution, and recycling of manufactured products as well as research and development. The digitization of manufacturing using fundamental cyber-physical production procedures and structures opens new opportunities. Hence, resources, including all workers, goods, systems, and procedures, should be combined as instances of adaptive, self-organized, cross-company, real-time systems. In addition, the human factor is now gradually being substituted with robots and complete automation in the sub-processes of implementation, surveillance, and control of procedures in the manufacturing sector, the leading industry in which the Industry 4.0 technique is implemented, particularly in the automotive and electronics industries. Robots and other machinery take over these operations functions. Here, the initial planning procedures are more significant than the elements of the conventional project management methodology. That is, it is necessary to restructure the recognition of the human component to consider its roles during the planning stage, execution, supervision, control, and completion. The two primary theoretical paradigms in project management, unpredictability and complexity, also play a significant role in IR 4.0 projects, modeling and specifying the attributes needed to oversee the workload and manage the escalating intricacy of the construction projects. According to Simion et al. (2018) (Simion et al., 2018), the automation of all operations, virtualization, globalization, standardization, and the shift from waterfall to agile are key features of project management in Industry 4.0, with more concentration on project-organizational relationships and the enhanced development of organizations in project management approach. As a result of the Industry 4.0 revolution, the key skills and competencies of project managers have also been transformed as compared to the skill requirements in traditional project management. These key competencies, shown in Figure 4.4, can be considered the major

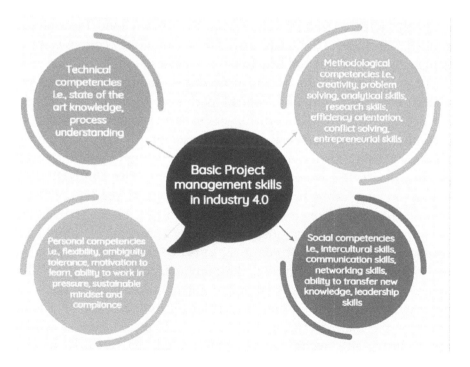

FIGURE 4.4 Industry 4.0 Related Basic Project Management Skills.

requirements for project managers to have. Without them, it will be very hard for anyone with traditional project management skills to thrive in their careers in the era of Industry 4.0.

EXAMPLE 4.1 THOUGHT BOX FOR STUDENTS

Students should concentrate on building specialized abilities that align with this technological revolution's distinctive possibilities and challenges if they want to flourish in project management within the larger context of Industry 4.0. Here are some crucial project management competencies for Industry 4.0 for students.

1. Technological literacy.
2. Data interpretation and assessment.
3. Technology flexibility.
4. Communication and cooperation.
5. Information risk and security management.
6. Strategic analysis.
7. Ethics-related matters.

4.2.3 Traditional vs. Industry 4.0 Project Manager's Competencies

In Industry 4.0 the required competencies have been transformed particularly related to the interactions with diverse stakeholders. Industry 4.0 places a lot of emphasis on real-time communication, which speeds up problem-solving and decision-making procedures. It also emphasizes management and knowledge sharing. To enhance the management of vital aspects and promote the growth of integrated collaboration and coordination, the data must be disseminated to all construction stakeholders. The importance of knowledge management in Industry 4.0 is like a gun in the hands of a skilled soldier. Without key skills and competencies relevant to Industry 4.0, project managers will have to face serious consequences in their projects and careers. Competency is defined as the mixed characteristics of personal skills and technical skills. It is instrumental to successfully, efficiently, and effectively complete the project. A comparison is presented in Figure 4.5 between traditional and IR 4.0 PM competencies. With the evolution of Industry 4.0, the skills and competencies needed for project managers have been transformed. In other words, it can be said that the traditional PM competencies have become obsolete in Industry 4.0.

	Traditional Project Managers	Industry 4.0 Project Managers
Project Constraints	Ensures adherence to budget, schedule, and scope.	Ensures adherence to budget, schedule, and scope.
Project Management Plan	Work with the Project Sponsor to develop the agreed Project Management Plan through interactions with stakeholders.	Work with the Project Sponsor to develop the agreed Project Management Plan through interactions with stakeholders.
Project Document Management	Execute and maintain project document management for the project cycle in compliance with the Authorities QA needs.	Execute and maintain project document management for the project cycle in compliance with the Authorities QA needs.
Accurate Reporting	Provide a timely and accurate report to the management team and team members.	Provide a timely and accurate report to the management team and team members.
Systems Engineering approach	Not present.	Ensure all project activities are undertaken in a systematic manner with a Systems Engineering approach at the core of the technical solution to project objectives.
Improvements through Deep Dive Analysis	Not present.	Provide deep dive analysis, proposal, and help implement improvements to a wide variety of cross organizational challenges.
Manage Projects in Real Time	Not present.	Manage project execution, risks; identify, resolve issues in a real time.
Complex systems execution	Not present.	Complex systems execution strategy formation and delivery.
Technical process Improvement	Not present.	Familiar with technical process improvement and development in start-up and complex settings.
Technical assessment of next generation software	Not present.	Drive the testing and technical assessment of next generation hardware and software technology with partners.
Manage cross-functional multi-location team member	Not present.	Motivate cross-functional, multi-location team members and manage deliverables to meet project milestones.

FIGURE 4.5   Traditional and Industry 4.0 PM Competencies.

Communication skills, leadership skills, and team-building skills have been transformed into real-time, automation, and simulation-based skills. The need for the incorporation of more advanced knowledge, that is, IoT, AI, automation, Cloud Computing, and Robotics has increased massively. Systems engineering approach, enhancement in projects through deep-dive analysis, management of projects in real-time, complex systems execution, technical process improvement, technical assessment of next-generation software and its incorporation in the construction projects, management of cross-functional multi-location teams are the major requisites or competencies needed for project managers. Industry 4.0 initiatives need a project manager with extensive expertise in the sector and a comprehensive understanding of CPS. Project team professionals or related virtual assistants are typically in charge of execution. Having experience with cutting-edge technologies and developments, predictive methods and algorithms, and big data analysis will likely help project managers concentrate on their task and manage it effectively so that it meets the clients' pre-established objectives. The abilities of project managers are founded on their vision to oversee the efficient operation of project management procedures to accomplish future projects not just in full but in every aspect. Users are always the center point in Industry 4.0, where they have the know-how and potential to use tools that lead to the project's success. This means that project managers must apply the following principles of lean and effective management. When integrating with devices, Industry 4.0 allows them to operate and undertake the project management principles effectively, which usually go through all areas of the firm and considerably help in smart or intelligent planning, manufacturing, production, and on-site execution.

4.2.3.1 Project Constraints

Project constraints are the drivers, factors, or variables that obstruct the smooth flow of the construction project. Time, cost, quality, scope, resources, safety, environment, and productivity are some of the major project constraints encountered by the stakeholders in their projects. When it is discussed in the context of traditional and IR 4.0 PM competencies, both focus on ensuring adherence to the time, schedule, and budget of the project.

4.2.3.2 Project Management Plan

The project management plan is a detailed document that incorporates all the management plans pertinent to the PMBOK knowledge areas. These plans play a crucial role in the successful execution of the project. Both traditional and IR 4.0 PM competencies focus on the coordination and collaboration with project stakeholders for the development of project management plans.

4.2.3.3 Accurate Reporting

Timely reporting about the project's progress to the stakeholders and top management is important for avoiding conflicts and disputes among team members, administrative staff, and key stakeholders. Both traditional and IR 4.0 PM competencies focus on efficient and timely reporting of the project to the project stakeholders.

4.2.3.4 Systems Engineering Approach

The systems engineering approach is a tried-and-true, methodical strategy that helps management by supporting management in precisely defining the objective or challenge, monitoring system functions and specifications, determining, and controlling risk, providing the foundations for sound decision-making, and confirming that products and services fulfill the requirements of customers. Both traditional PM competencies lack this important area and do not cover this aspect. IR 4.0 PM competencies comprehensively focus on this aspect by coordinating and collaborating with all the project components in a systematic way.

4.2.3.5 Improvement through Deep-Dive Analysis

According to one definition, a deep-dive evaluation is "a strategy of swiftly engaging a team into a situation, to offer solutions or generate ideas." Deep-dive analyses typically concentrate on things like process, structure, governance, and culture. Traditional PM competencies lack this area, whereas IR 4.0 PM competencies cover this aspect by carefully analyzing the cross-organizational operations, functions, and challenges. Moreover, solutions for resolving the highlighted challenges are also provided through deep-dive analysis.

4.2.3.6 Manage Projects in Real Time

Management of projects in real time means that the PM is constantly obtaining data on the project in real time and updating it. The issues originating from the operations of the project are resolved through the real-time data obtained from the projects. Moreover, stakeholders are kept informed about the real update of the project, and their concerns are resolved well in time. This practice saves time and assists the PM in effectively executing the project and managing the encountered risks and challenges.

4.2.3.7 Complex Systems Execution

Executing complicated systems, which are composed of interconnected elements that work together with one another to accomplish a given goal, is referred to as complex systems execution. IR 4.0 PM competencies focus on this aspect because, with each passing day, construction projects are becoming more complex with the interdependence of a large number of tasks, activities, processes, procedures, methodologies, stakeholders, and public and private entities. Through complex systems execution, the complex nature of the construction projects can be managed proficiently in contrast to traditional PM.

4.2.3.8 Technical Process Improvement

Process enhancement entails enhancing both the business it supports and the internal management system's procedures. Finding and implementing adjustments to processes that will raise the caliber, efficiency, and productivity of software development is the goal of technical process improvement. The world is transforming towards digitization; therefore, the inclusion of technical process improvement through real-time data is crucial. IR 4.0 PM competencies focus on this key aspect by constantly improving the complex processes and procedures of undertaking business activities with the incorporation of modern technologies pertinent to IR 4.0.

4.2.3.9 Technical Assessment of Next-Generation Software

Next-generation software is developed by incorporating the latest technology and tools such as AI, machine learning, and deep learning. Through technical assessment of these next-generation software, the organizations can lead the markets, test and modify the software and hardware requirements, and win the trust of their sponsors. This is one of the major requirements of IR 4.0 PM competencies.

4.2.3.10 Manage Cross-Functional Multi-Location Team Members

When many teams within an organization work together on a project or a goal in common, this is known as cross-functional collaboration. For organizations trying to eliminate barriers and work with departments rather than against them, collaboration is a crucial practice. Workplaces that encourage collaboration and management of cross-functional multi-location team members are both more productive and more enjoyable.

4.2.4 Intelligent and Smart Manufacturing

Furthermore, in Industry 4.0, intelligent manufacturing or smart manufacturing is a concept with a wider scope in which the key aim of manufacturing lies in optimizing production and product trans-actions with the incorporation of advanced high-tech information and manufacturing tools and techniques. In construction projects, smart manufacturing and execution can play a key role in which, through the utilization of advanced tools and techniques, the integration of production and procurement of materials, quality assurance and quality control, optimized processes, and maintenance data can be stored, managed, and controlled from a single platform, as shown in Figure 4.6. This can be done in real-time because the data stored

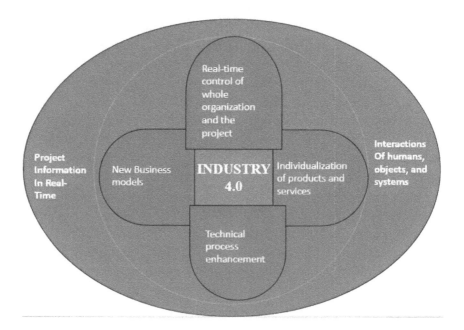

FIGURE 4.6   Smart and Intelligent Management of Projects in Industry 4.0.

can be shared with concerned stakeholders, and their shared data can be received efficiently. Management and integration of project data from a single database help in improved data sharing between involved stakeholders, enhanced data security, better data integration, improved data reach and decision-making, and optimized productivity with real-time tracking and monitoring of productivity.

Furthermore, the concept of intelligent, or smart, manufacturing has been around in some form or another since the early nineteenth century. This type of engineering seeks to streamline and automate production by using advanced technologies such as Robotics, AI, and IoT. At its core, intelligent manufacturing centers on using a combination of human expertise, sophisticated algorithms, and intelligent machines to maximize production efficiency. This approach seeks to achieve better outcomes with fewer resources and automation of manual labor and should result in quality products produced faster and cheaper than traditional methods. The key elements of intelligent manufacturing are the development and use of various technologies and associated processes. One of the most widely used technologies is AI, which helps with decision-making and predictive analytics, based on data supplied by the manufacturing process. Automated machines can work together in close cooperation, with no physical interaction and no human intervention, to produce parts and components through Cloud Computing. The robots should also be able to self-monitor and adjust operations, to ensure that the most efficient production methods are always being used.

The next important element is the IoT, which provides the essential link between machines in the factory and factories in distant places. This connectivity allows manufacturers to control the flow of materials and products, create new orders, monitor machine performance, and track temperatures and other environmental factors. Further, it can serve as a communication and control channel between operators and machines, allowing them to quickly respond to any issues. Also, intelligent manufacturing systems should be able to operate in a continuous learning loop, which can further improve the system's performance over time. It should also be able to capture data from its working environment, to make sure that the manufacturing process is compliant with industry standards. This ensures that all processes, from design to production to delivery, are always running efficiently. These technologies and processes, when used together, provide a comprehensive system for intelligent and smart manufacturing. By taking advantage of the latest advancements in technology and processes, the approach provides the opportunity for companies to increase the quality and quantity of the products they produce while staying ahead of the competition. With increased production efficiency, manufacturers can focus more on improving the customer experience by providing better service and staying ahead of trends.

4.2.4.1 New Business Models

The rule of a successful business is the consistent modification of the business models or the development of new business models with changing technological landscapes. Figure 4.7 highlights the incorporation of certain crucial information before the development of new business models. Two tiers of information should be focused on effective business model development such as organizational response and addressing strategic challenges and opportunities. After the identification of strategic challenges and opportunities in the context of new business, this information has collaborated with the organizational response. Inputs are provided, and analysis is done with this information. The outcome is the business model that incorporates the resource streams and expense or cost streams.

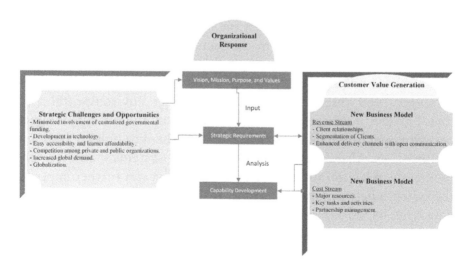

FIGURE 4.7   New Business Model Development Framework.

4.2.4.2 Technical Process Enhancement

To improve both productivity and quality, get rid of operational bottlenecks, and reduce expenses, process optimization aims to find and fix weak points in the company processes. Organizations that do not update their processes with the changing requirements in the business world die with their obsolete processes in a short time. Figure 4.8 shows a cyclic process of business technical process enhancement. It should undergo plan, do, check, and act phases, and all the necessary inputs should be made part of it. Moreover, along with input variables, customer feedback should also be included in this process.

4.2.4.3 Individualization or Personalization of Products and Services

Customizing products entails altering them to better meet the tastes and needs of clients. This could either be custom products created for a specific customer or built-to-order products using a standardized production process with a range of options accessible within current constraints. Individualization in marketing is the process of tailoring a product or service to each customer. Customizing the way consumers interact with unique content, goods, and promotions can help in the attainment of intelligent manufacturing.

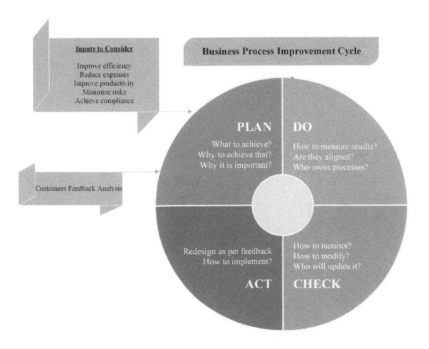

FIGURE 4.8 Business Process Improvement Cycle.

4.2.4.4 Real-time Control of the Whole Organization and the Project

The capacity to monitor and manage every part of an organization or project in real time is referred to as "real-time control of the entire organization and the project." This entails keeping an eye on key performance indicators, spotting potential hazards, and implementing the necessary changes to ensure the project or organization is on track to achieve its objectives. Real-time control for organizations and projects should comprise definitions of objectives, collection of real-time data, processing of data, and analysis and visualization of the data, as shown in Figure 4.9. This should then be used to update the existing parameters related to that organization or the project.

> **EXAMPLE 4.2 THOUGHT BOX FOR STAKEHOLDERS**
> - By implementing cutting-edge technology and procedures, the construction sector may accomplish intelligent and smart production, often known as smart construction or digital construction.
> - Construction sector players, such as contractors, architects, engineers, technology suppliers, and regulatory agencies, must work together to establish an innovation culture, invest in research and development, and achieve intelligent and smart manufacturing. Smart building practices will also be successfully implemented if the appropriate skill sets, along with a knowledge base, are developed through instructional and upskilling programs.

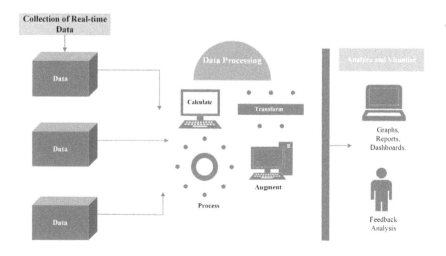

FIGURE 4.9   Real-Time Control Framework.

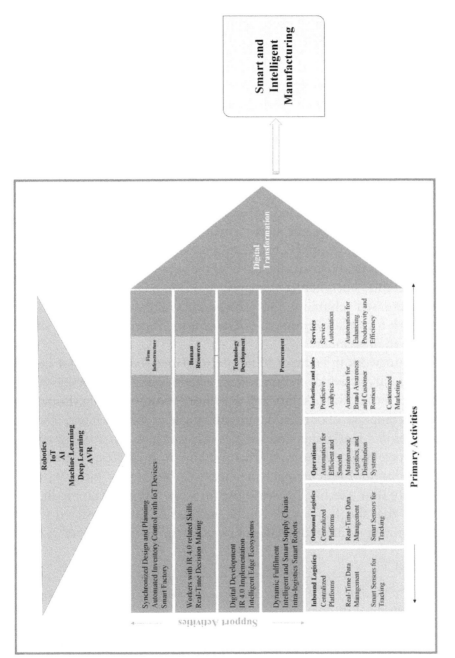

FIGURE 4.10 Smart and Intelligent Manufacturing Framework for Organizations.

4.2.5 Intelligent and Smart Supply Chain Management

Modern technology must be used to optimize various aspects of the supply chain to practice intelligent supply chain management. Companies can better optimize their supply chains by using an intelligent supply chain, which makes use of contemporary technologies for data collection and insight development. This can help them cut expenses, boost profitability, deliver customers more quickly, and outperform their competition. When companies fully utilize the most recent innovations in retail digital transformation technology, including artificial intelligence (AI), machine learning (ML), predictive analytics, integrated commerce, and big data, intelligent supply chain management is conceivable.

Business intelligence for the supply chain is being fueled by artificial intelligence, which includes machine learning and predictive analytics. These technological advancements enrich and enhance every aspect of the digital supply chain using big data, logistics trends, and business patterns. Historically, parts of the supply chain were manual and unreliable, but intelligent supply chain management has significantly increased reliability and transparency. Figure 4.10 presents a framework for organizations to implement intelligent and smart supply chain management.

The framework in Figure 4.10 has been divided into two major tiers, that is, primary activities and supporting activities tier. Both complement each other in attaining smart and intelligent manufacturing. Primary activities comprise inbound and outbound logistics management, operations, sales and marketing, and services management, whereas supporting activities include firm infrastructure, human resources, technology development, and procurement management. Both tiers integrate the adoption of IR 4.0 technologies to close the loop. It is important to highlight here that this framework can be adopted by any industry if they want to achieve smart and intelligent manufacturing in their operations.

REFERENCES

PMI. 2017. *A Guide to the Project Management Body of Knowledge*. Project Management Institue.

Simion, C.-P., Popa, S.-C. & Albu, C. 2018. Project management 4.0–project management in the digital era. *12th International Management Conference*. Editura ASE, 93–100.

CHAPTER 5

IR 4.0 Integration with PMBOK and Sustainability

5.1. CONCEPT OF SUSTAINABILITY

Over the past few decades, the idea of sustainability has gained prominence in discussions and actions on a global scale. It signifies a profound change in how we view the interdependence of human society and the natural world. Sustainability is not merely a catchphrase or a fashion; it is a vital necessity for the sustainability of our planet and the welfare of future generations. Sustainability is fundamentally the ability to endure and endure over time. It refers to the notion that we must fulfill our wants now without endangering the capacity of future generations (Larsson and Larsson, 2020). When the Brundtland Report was released in 1987, it gave this idea more recognition by defining it as "development that meets the needs of the present without compromising the ability of future generations to meet their own needs (Khurshid et al., 2023). Three interconnected pillars of sustainability, each of which is crucial for creating harmony and balance in our environment, are shown in Figure 5.1.

It includes initiatives to lessen pollution, preserve biodiversity, combat climate change, and safeguard the planet's health. Environmental sustainability acknowledges that, because we are part of a complex

DOI: 10.1201/9781032621760-5

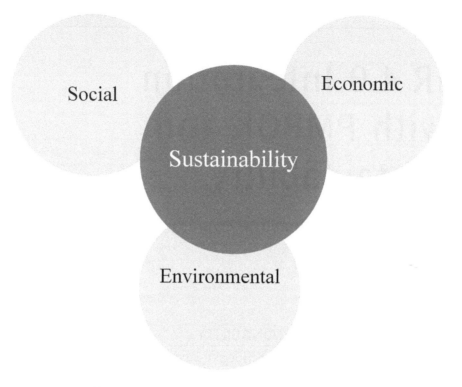

FIGURE 5.1   Pillars of Sustainability.

web of life, the health of the ecosystems on earth affects us directly (Ghorbani et al., 2022). Social sustainability emphasizes the equity and well-being of people. It entails building diverse, equitable, and resilient communities where everyone can benefit from opportunities, access to necessary resources, and a good standard of living. This pillar addresses problems including social justice, healthcare, education, poverty, and cultural diversity (Piyathanavong et al., 2022). Economic sustainability aims to maintain stable, effective, and fair economic systems. Responsible resource management, ethical business conduct, and the pursuit of economic progress without endangering the environment or taking advantage of disadvantaged groups are all part of it. Economic sustainability considers the fact that affluence is a means to an end rather than an end in and of itself (Vásquez et al., 2021).

Social sustainability is critical to raising everyone's standard of living, regardless of their circumstances or background. It aims to end poverty, lessen inequality, and advance access to social services, healthcare, and education. Building strong, resilient economies that can endure shocks and disruptions is the goal of economic sustainability (Adriana and Ioana-Maria, 2013). Sustainability represents a moral and ethical obligation to take care of the environment and make sure that our actions don't affect other people, including current and future generations. Although the idea of sustainability is appealing, it faces numerous difficulties and barriers:

Thinking short-term: In many civilizations, short-term political and economic gains are given precedence over long-term sustainability (Obradović et al., 2019). This may result in choices that put the interests of present generations ahead of those of future ones. Overexploitation of resources causes stress on the planet's ecosystems and the loss of essential resources like fresh water, arable land, and fossil fuels. Given that sustainability necessitates long-term planning and thinking, it can be contradictory to think about sustainability in the short term. However, there are circumstances in which a short-term outlook might be crucial to accomplishing sustainability objectives.

To solve urgent environmental, social, or economic concerns without compromising long-term sustainability goals, short-term sustainability refers to actions and initiatives performed right away. While long-term planning is essential for protecting the welfare of future generations and the planet, there are some circumstances where immediate action is required to avert irreparable harm. Crisis management in the aftermath of natural disasters is an illustration of short-term sustainability.

Social injustice and inequality: The quest for social sustainability is hampered by the persistence of social injustice in many regions of the world. achieving social and equity. Social injustice and inequality are rooted in societal problems that sustain structural gaps in access to resources, opportunities, and fundamental rights. They affect marginalized communities negatively on several different levels, including racial, gender, and economic status. These disparities not only limit individual potential but also threaten the values of justice and fairness that are the foundation of societies. Dismantling oppressive structures, advancing equity, and establishing inclusive environments where everyone may thrive regardless of their background or identity are necessary to address social injustice and inequality. It is a moral requirement and a crucial step in making the world more just and equal for all people.

Political and economic interests: It can be challenging to implement sustainable policies and practices since strong political and economic interests frequently oppose changes that could affect their hegemony or profit margins (Banihashemi et al., 2017). Politicians frequently have concerns regarding the general public's perception and electoral support when it comes to sustainability. Politicians may place a higher priority on issues that appeal to people, including promoting renewable energy programs or preserving national parks. The adoption of particular sustainability goals and targets may result from political parties using sustainability as a platform to set themselves apart from their rivals. Global sustainability initiatives also heavily depend on international politics. Countries negotiate their obligations to solve global concerns in agreements like the Paris Agreement on climate change. These agreements show a willingness to protect the environment as well as an understanding of the financial benefits that come with using sustainable energy sources.

Despite these obstacles, there are many approaches to achieve a sustainable future:

Education and awareness: Promoting a culture of sustainability can be done by educating people about sustainability from a young age. Governments have the power to adopt laws and regulations that reward sustainable behaviors and punish those that are not.

Collaboration and partnerships: Governments, corporations, civic society, and academics can work together across sectors to advance sustainable development (Armenia et al., 2019). To address complex and interconnected difficulties, collaboration and partnerships are essential building blocks of sustainability efforts (García-Gómez et al., 2021). They promote group action and tap into the strengths of many stakeholders. With its comprehensive character spanning environmental, social, and economic components, sustainability frequently necessitates the joint efforts of governments, corporations, nonprofit groups, and communities. Cooperation and collaborations are tools for combining resources, transferring knowledge, and enhancing effect.

Consumer preferences: Consumer preferences and the demand for environmentally friendly goods and services might encourage companies to use more environmentally friendly practices. Consumer preferences are the specific options and considerations that one or more consumers have while making a purchase. Numerous elements, such as individual beliefs, preferences, and requirements, as well as cultural and economic forces, influence these inclinations. The ability to adjust products, services, and policies to the needs and expectations of their target audience depends on firms and policymakers being aware of customer preference.

Local initiatives: The implementation of sustainable practices at the community level depends heavily on grassroots movements and local initiatives. Initiatives that are taken locally or regionally to solve problems or foster good change are referred to as local projects, programs, or activities. These programs are frequently the result of grassroots efforts, community organizations, local governments, or concerned citizens who have a strong sense of responsibility for the needs and problems in their immediate surroundings. Local projects can encompass a wide range of sectors, including social, environmental, economic, and cultural aspects.

Sustainability is not merely a catchphrase; rather, it is a fundamental paradigm change that necessitates adopting a new way of thinking and behaving. It is a comprehensive strategy that acknowledges the interdependence of social, economic, and environmental well-being. Although achieving sustainability involves overcoming many difficulties and hindrances, the need to do so is more urgent than ever. Sustainability is a moral, ethical, and political imperative (Rezghdeh and Shokouhyar, 2020).

Sustainability has recently taken the front stage in project management, revolutionizing how tasks are organized, carried out, and assessed. Sustainability in project management includes social, economic, and environmental factors in addition to the conventional emphasis on completing projects on schedule and within budget (Aarseth et al., 2017). The importance of sustainability in project management is significant and multidimensional, having an impact not only on the overall success of society and the environment but also on the success of specific projects. The pressing necessity for sustainability in project management is one of the main reasons for its importance (Qureshi et al., 2022).

The urgent need to solve environmental concerns is one of the main reasons for the importance of sustainability in project management. Organizations must now lower their ecological footprints because of climate change, resource scarcity, and environmental damage. Energy efficiency, waste minimization, and ethical resource management are just a few examples of practices that sustainable project management promotes. Organizations aid in the preservation of ecosystems and the reduction of climate change by incorporating environmental considerations into project design and execution.

Social responsibility is a component of project sustainability. The well-being of communities, workers, and stakeholders can be significantly impacted by projects' deep social effects. Sustainable project management emphasizes moral behavior, social equality, and participation in society considering this (Lance, 2022). It entails taking the needs and concerns into account. It entails considering the requirements and worries of diverse stakeholders, particularly underrepresented groups, and making sure that projects have a beneficial social impact. Project managers help create inclusive and equitable communities by putting social responsibility first.

Sustainability in project management includes considerations for the economy as well as the environment and society. Projects need to be financially feasible, add value to organizations, and reduce waste and inefficiency. The use of cost-effective methods, ethical resource allocation, and long-term economic viability are all promoted by sustainable project management. It aids businesses in maximizing resource usage, lowering operating expenses, and improving financial performance, all of which support economic sustainability (Hu et al., 2022).

Environmental impact assessments and other sustainability-related laws and regulations are very strict in many places. By using sustainable project management, projects that use sustainable project management are guaranteed to follow these rules. It entails carrying out environmental impact analyses, securing required licenses, and upholding sustainability standards. Compliance not only indicates an organization's commitment to ethical business practices but also aids in avoiding legal and regulatory problems (Ibrahim et al., 2021).

An organization's reputation is improved, and its relationships with stakeholders are strengthened via sustainable project management. Customers, investors, and the general public have higher expectations of businesses in terms of their ethical behavior and commitment to sustainable development. Stakeholders have a more positive opinion of projects that adhere to sustainability principles. A good reputation can foster higher client loyalty, investor confidence, and better linkages with the community, all of which can be essential for the success of upcoming projects (Mortaheb and Mahpour, 2016). A skilled staff with a thorough awareness of sustainability's ideas and practices is necessary for sustainability in project management. Organizations and academic institutions must make investments in capacity building and train project managers to give them the skills and resources they need to manage sustainable projects successfully.

Project management sustainability requires being proactive. Regulatory changes, reputational risks, supply chain disruptions, and environmental emergencies are just a few examples of the sustainability-related risks that it aids organizations in identifying and assessing (Leoto, 2020). Organizations can create ways to manage these risks and guarantee project success via early risk identification. In a world that is getting more complicated and uncertain, sustainability-driven risk management strengthens the resilience of projects and organizations. Determining, measuring, and reporting sustainability outcomes is one of the key difficulties in the sustainability-project management nexus. Key performance indicators (KPIs) that reflect the project's effects on the environment, society, and economy must be established by project management. Accountability and proving a project's contribution to more general sustainability goals depend on transparent reporting on sustainability measures.

Innovation is encouraged by sustainable project management, which also creates new market prospects. Businesses that uphold sustainability ideals frequently come up with novel answers to difficult problems. Projects centered on sustainable agriculture, green building, or renewable energy, for instance, might take advantage of the expanding markets for eco-friendly goods and services. Organizations establish themselves as leaders in new markets and gain a competitive edge by coordinating projects with sustainability trends (Unterhofer et al., 2021). Collaboration among many stakeholders, including governmental and nongovernmental organizations, local communities, and business partners, is frequently necessary for sustainability efforts. To successfully traverse the complicated world of sustainability expectations and standards, project managers must excel in stakeholder engagement and collaboration. Building partnerships and engaging in effective communication are essential management abilities for sustainable project management. Determining, measuring, and reporting sustainability outcomes is one of the key problems in the relationship between sustainability and project management. Key performance indicators (KPIs) that reflect the project's effects on the environment, society, and economy must be established by project management. Accountability and proving a project's contribution to more general sustainability goals depend on transparent reporting on sustainability measures. The long-term perspective is encouraged by sustainability in project management.

Organizations consider the long-term effects of their projects on the environment, society, and economy rather than just concentrating on short-term advantages (Silvius, 2017). For initiatives to have lasting value and ensure that they benefit future generations, they must have this long-term perspective. It is impossible to stress the importance of sustainability in project management (Aalavi and Janatyan, 2020). By including environmental protection, social responsibility, economic viability, regulatory compliance, reputation development, risk mitigation, innovation, and long-term vision, it goes beyond the conventional definitions of project success (Hashemi-Tabatabaei et al., 2022). Projects that are managed sustainably link themselves with sustainability's overarching objectives, making the world more just, resilient, and responsible.

5.2. SUSTAINABILITY AND PROJECT MANAGEMENT

The idea of sustainability has become increasingly important in recent years as the world community battles a variety of environmental, social, and economic issues. Project management and other decision-making processes now prioritize sustainability due to the urgency of addressing climate change, resource depletion, social injustices, and economic imbalances (Silvius and Schipper, 2016). Sustainability, often known as "sustainable development," is the pursuit of behaviors and methods that satisfy present demands without jeopardizing the capacity of future generations to satisfy their own needs (Walters et al., 2018). The discipline of project management involves organizing, planning, and carrying out projects to accomplish certain goals and objectives. It includes several things, including stakeholder communication, resource allocation, and scope definition, as shown in Figure 5.2. In the past, project management mostly concentrated on completing projects on schedule and on budget (Fernández, 2019). Project management has changed to include sustainability issues as

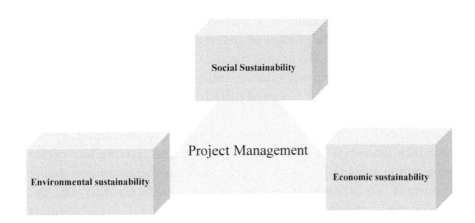

FIGURE 5.2   Project Management and Sustainability.

problems with sustainability have grown more serious (Silvius and Schipper, 2020). Setting sustainable project objectives is one of the main strategies to incorporate sustainability into project management. These aims connect the project's objectives to more general sustainability goals. For instance, a construction project might decide to employ sustainable building materials, decrease trash, or minimize energy use (Herrera-Reyes et al., 2018). These goals act as a set of guiding principles for the duration of the project.

Life cycle assessment (LCA) approaches are being used more frequently by project managers to examine the environmental impact of projects from inception to conclusion. LCAs consider a project's energy, material, and process footprints as well as its overall environmental impact. Project teams can find ways to cut emissions, use resources less, and make more sustainable decisions by doing LCAs (Buhulaiga et al., 2019).

Beyond environmental factors, sustainability in project management encompasses many other factors. Both social and economic factors are present. It is crucial to involve all relevant parties, including employees, marginalized groups, and local communities. The objectives of projects should be to stimulate community growth, pay fair salaries, and provide safe working conditions. Resource effectiveness is emphasized in sustainable project management. This covers the effective utilization of raw materials, water, and energy (Sudhir and RC, 2023). Projects can use water conservation and energy-saving technologies. Project outcomes may be impacted by sustainability risks such as legislative modifications, reputational hazards, and supply chain disruptions. To ensure the success of their projects, project managers must recognize, evaluate, and reduce these risks. To foresee and address sustainability-related difficulties, proactive risk management is a necessary component of sustainable project management. It is essential to regularly monitor and report on sustainability performance (Stanitsas et al., 2021). Tracking and sharing key performance indicators (KPIs) with stakeholders on sustainability goals is important. Transparent reporting improves accountability and permits changes while a project is being carried out.

There are many advantages to incorporate sustainability into project management, but there are also difficulties (Ohueri et al., 2023). These difficulties include the requirement for specialized knowledge, opposition to change, and increased expenditures related to sustainable practices. The opportunities, though, are substantial. Frequently, sustainable projects, greater stakeholder interactions, lower operational costs, greater reputation, and more access to markets with sustainability standards are frequent benefits of sustainable initiatives. In the modern environment, sustainability and project management are inextricably interwoven (Rezahoseini et al., 2019). By incorporating sustainability principles into project planning, implementation, and evaluation, project managers may play a crucial role in tackling the issues of global sustainability.

Project managers can support the overarching objective of achieving harmonious economic, environmental, and social development by establishing sustainable project objectives, conducting life cycle assessments, engaging stakeholders, promoting resource efficiency, assessing and mitigating risks, and monitoring performance (Co kun, 2019). In the contemporary world, sustainability is not a choice but a requirement, and project management is a powerful tool for fostering good change towards a more sustainable future. Project management that prioritizes sustainability must go beyond the current stages of the project. A tool called life cycle assessment (LCA) is used to examine how a project's or product's environmental effects change throughout the course of its full life cycle, from the extraction of raw materials to disposal. To find chances for lowering environmental footprints, boosting efficiency, and enhancing sustainability, project managers must integrate LCA and design thinking into their operations.

5.3. IR 4.0 AND ECONOMIC SUSTAINABILITY 4.0

The Fourth Industrial Revolution (IR 4.0), which altered industries, business structures, and the very nature of employment, has caused a seismic change in the world economy. Fundamentally, IR 4.0 is the convergence of digital technologies, AI, the Internet of Things (IoT), big data analytics, and automation, ushering in a period of unprecedented connection and productivity shown in Figure 5.3. However, IR 4.0's influence goes well beyond technological developments; it also has significant effects on the viability of the economy (Marcelino-Sádaba et al., 2015). In the context of IR 4.0, economic sustainability is a multidimensional notion that includes both traditional economic development pillars like growth and profitability as well as broader issues like environmental and social responsibility. It aims to establish a careful balance between economic advancement and social growth.

The potential of IR 4.0 to improve resource efficiency is one of the fundamental features that support economic sustainability. Industries

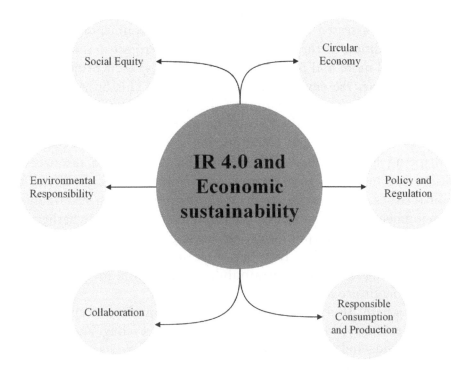

FIGURE 5.3   IR 4.0 and Economic Sustainability.

can optimize their operations, cut waste, and lessen their environmental impact with modern data analytics and IoT. For instance, smart factories can modify their manufacturing procedures in response to changes in demand in real time, which reduces energy use and greenhouse gas emissions. Such resource optimization not only benefits the environment but also lowers costs for companies, increasing their long-term economic viability (Martens and Carvalho, 2017). Additionally, IR 4.0 encourages the creation of sustainable goods and services. There is a rising need for environmentally friendly and socially responsible products as customers' awareness of environmental and social issues grow. The necessity for long-distance transportation and the related carbon emissions is diminished using technologies like 3D printing, which facilitate more effective and localized manufacturing. Blockchain technology may also create transparent supply chains, guaranteeing that goods are manufactured and sourced ethically, adding to economic sustainability by reflecting customer values (Carvalho and Rabechini Jr, 2017). The triple bottom line approach is a key idea at the junction of sustainability and project management. With this method, projects are assessed for their social and environmental implications in addition to their financial viability. Project managers are being urged more and more to think about how their projects will impact people, the environment, and revenue. This broader viewpoint necessitates a reevaluation of the goals and trade-offs of the project, which frequently results in more sustainable decision-making. Partnerships and collaboration are crucial sustainability-enhancing factors. They make it possible to combine resources, skills, and initiatives from various stakeholders to tackle difficult global issues. Partnerships are the means through which sustainable development becomes a shared duty and a joint success, whether at the global level, in corporations, among civil society organizations, or in local communities. Together, we can more successfully negotiate the complex web of sustainability and build a more just and sustainable future for all.

Additionally, IR 4.0 equips companies to adopt the concepts of the circular economy, which encourages the reuse, recycling, and renovation of goods and resources. IoT sensors can follow a product's life cycle, allowing businesses to spot opportunities for recycling parts or materials, cutting waste, and extending the life of products (Silvius et al., 2017). There is a rising need for environmentally friendly and socially responsible products as customers' awareness of environmental and social issues grows (Silvius et al., 2012). The necessity for long-distance transportation and the related carbon emissions is diminished using technologies like 3D printing, which facilitates more effective and localized manufacturing. Businesses that use circular economy practices not only lessen their environmental effect but also create new revenue streams through remanufacturing and product-as-a-service models. IR 4.0 can enhance working conditions and produce inclusive economic opportunities in social responsibility. Robotics and automation can handle risky and boring activities, lowering workplace accidents and raising general worker satisfaction (Martínez-Perales et al., 2018). Additionally, programs for upskilling and reskilling the workforce can ensure that nobody is left behind by preparing them for the jobs of the future. This attention to employee welfare and fair access to technology is consistent with the tenets of economic sustainability, which hold that economic growth should benefit society (Yazici, 2020).

It's important to recognize that not everyone will gain equality from IR 4.0. Automation and AI run the risk of displacing workers in specific industries, which could worsen income inequality. Thus, to ensure a fair transition to the digital economy, governments and corporations must develop laws and plans (Silvius and Schipper, 2014). The idea of economic sustainability can be strengthened by programs like universal basic income, lifelong learning opportunities, and job placement aid that can help to buffer the adverse social effects of automation (Yazici, 2020).

By increasing resource efficiency, promoting sustainable goods and services, encouraging circular economy practices, fostering better working conditions, and providing inclusive economic possibilities, the Fourth Industrial Revolution presents unheard-of chances to improve economic sustainability. To ensure that the advantages of IR 4.0 are utilized in a way that is consistent with the more general objectives of economic, environmental, and social sustainability, governments, enterprises, and individuals must work together to ensure that these benefits are realized (Dreyer et al., 2020). The secret to creating a future in which prosperity is shared by all and the planet is preserved for future generations is to balance technological growth with responsible and ecological practices.

5.4. IR 4.0 AND ENVIRONMENTAL SUSTAINABILITY 4.0

The Fourth Industrial Revolution (IR 4.0), which is being fueled by developments in digital technology, artificial intelligence, automation, and networking, marks a major transformation in the way we live and work. The globe is simultaneously dealing with a critical dilemma related to environmental sustainability. The loss of biodiversity, resource depletion, and climate change are issues being faced by the entire world (Poveda-Orjuela et al., 2019). Nevertheless for environmental sustainability attainment, IR 4.0 carries certain factors in this context, which holds the benefits and risks, thus, a comprehensive understanding of their interaction is essential for creating a sustainable future as shown in Figure 5.4. Sustainability efforts are crucially financed by financial entities, such as banks, investment funds, and charitable organizations. Financial institutions and organizations that are concerned with sustainability can work together to allocate funding towards initiatives that are both socially and environmentally responsible. For instance, impact investment invests in projects that produce both financial returns and advantageous social or environmental benefits. Even if partnerships and collaboration in sustainability have many advantages, there are still difficulties. Stakeholders' divergent goals, priorities, and values may make collaboration difficult. Obstacles could also be communication breakdowns and problems with transparency and trust. Clarified goals, mutual trust, and the assurance that all partners are sincerely devoted to the sustainable cause are essential for overcoming these obstacles.

IR 4.0 can support environmental sustainability in several ways, including by maximizing resource usage. Real-time monitoring is made possible by the digitalization of industries, which is made possible by innovations like the Internet of Things (IoT) and big data analytics. Smart energy grids, for instance, can dynamically balance supply and demand, lowering energy waste and greenhouse gas emissions. Like this, precision agriculture makes the best use of data-driven insights to minimize the use of water, fertilizers, and pesticides while improving yields to meet the needs of a growing world population (Chawla et al., 2018). The long-term viability and stability of economic systems is emphasized as an essential element of overall sustainability. It comprises managing resources, investments, and economic activities in a way that meets present needs while

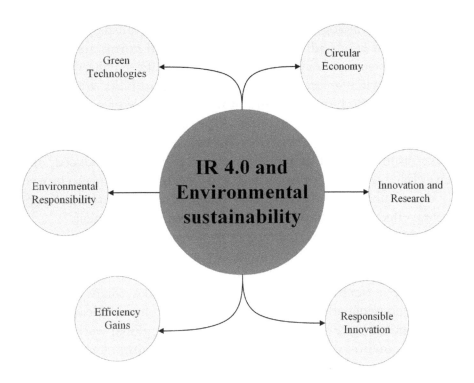

FIGURE 5.4   IR 4.0 and Environmental Sustainability.

preserving the well-being and financial security of future generations. The idea of responsible resource management, where resources
are utilized effectively, waste is eliminated, and adverse environmental effects are minimized, is essential to economic sustainability. This
strategy ensures that natural resources will be available for economic
activity in the long run by preserving them for future generations and
preventing the depletion of scarce resources. The encouragement of
inclusive economic growth and shared prosperity is a crucial component of economic sustainability. Sustainable economies strive to
ensure equal access and lessen economic disparity. Additionally, IR
4.0 supports the creation of sustainable goods and services.

Customers are becoming more environmentally concerned and are
demanding products that are ethically sourced and environmentally
friendly. Local and on-demand manufacturing is made possible by IR
4.0 technologies like 3D printing, which lowers the environmental
impact of long-distance shipping and surplus inventories. Additionally,

blockchain technology can improve supply chain transparency, guaranteeing that goods adhere to moral and ethical norms and so supporting the tenets of environmental sustainability. Furthermore, the adoption of circular economy principles is a possibility with IR 4.0 (Yu et al., 2018). Industries can follow the life cycle of products and materials, discovering opportunities for reuse and recycling, by utilizing IoT sensors and data analytics. By switching from a linear "take-make-dispose" model to a circular one, less waste is produced, and fewer finite resources are extracted. Products that are made to be easily disassembled and recycled further reduce their negative environmental effects and support sustainable resource management.

IR 4.0 can accelerate the switch to greener and more renewable energy sources. More effective energy grids are made possible by modern monitoring and control systems, which can accommodate more intermittent renewable energy sources like solar and wind power. Consequently, IR 4.0 opens the door for a decarbonized energy industry, which is necessary for reducing global warming and ensuring environmental sustainability. The environmental effects of IR 4.0, however, are not without difficulties. Massive amounts of energy are consumed by the proliferation of digital devices and data centers, which increases carbon emissions. Concerns regarding resource depletion and environmental contamination are also brought up by the production of electronics and the disposal of e-waste. To reduce these negative effects, it is essential to develop energy-efficient technologies, appropriately recycle electronic trash, and encourage sustainable design and production methods (Marnewick, 2017). In areas that are largely dependent on traditional employment sectors, the widespread adoption of automation and artificial intelligence in industry may cause job dislocation and social instability. As displaced employees, this relocation may make environmental problems worse. To ensure that the workforce can adjust to the shifting employment landscape while keeping an eye on environmental sustainability, a just transition combining reskilling and upskilling programs is essential. As a result, opportunities and difficulties presented by IR 4.0 and environmental sustainability 4.0 are intricately entwined. The digital revolution provides strong instruments for promoting sustainable goods and services, embracing the ideas of the circular economy, and switching to clean energy sources. However, there are concerns associated with energy use, e-waste, and social repercussions. In the era of IR 4.0, harnessing its promise while mitigating its negative externalities calls for a deliberate and coordinated effort. Society may benefit from IR 4.0 by embracing innovation, regulation, and ethical business practices.

5.5 IR 4.0 AND SOCIAL SUSTAINABILITY 4.0

The Fourth Industrial Revolution (IR 4.0) has brought about an era of unheard-of technological advancement, transforming economies, cultures, and how we connect with the outside world. The significance of social sustainability, which entails promoting inclusive, equitable, and just communities that meet the needs of the present without jeopardizing the welfare of future generations, is also becoming more widely acknowledged, as shown in Figure 5.5. With its potential for transformation, IR 4.0 offers both opportunities and difficulties for social sustainability in the twenty-first century. A key component of overall sustainability is social sustainability, which emphasizes the welfare, equity, and social cohesiveness of communities and societies. It includes a wide range of elements and aspects that work to build inclusive, robust, and prosperous communities for both the present and the next generations. Fundamentally, social sustainability acknowledges that resolving social challenges and advancing social fairness are necessary for achieving economic and environmental sustainability. Making sure that every member of a community has access to fundamental human rights and opportunities is a crucial component of social sustainability. This includes having access to jobs, housing, clean water, healthcare, education, and other necessities. To incorporate vulnerable or marginalized groups into society's fabric and actively combat inequality and prejudice, social sustainability also requires developing social inclusion.

The possibility of raising living standards and eliminating social injustices is one of the most important opportunities provided by IR 4.0. Artificial intelligence (AI) and automation technologies have the potential to boost productivity and open new business prospects. However, these advantages must be shared fairly to promote social sustainability. To prevent automation and AI from widening wealth gaps rather than promoting more economic inclusion, governments, corporations, and civil society must collaborate.

FIGURE 5.5   IR 4.0 and Social Sustainability.

Proactive worker development and reskilling initiatives are one approach to achieve this. Reskilling programs can assist people in transferring new tasks that need expertise in fields like data analysis, digital marketing, and programming when automation replaces some job roles. Additionally, IR 4.0 makes remote learning and online learning platforms possible, increasing the accessibility of education and skill-building for underserved areas. This may enable people to engage in the digital economy and support social sustainability. Additionally, IR 4.0 encourages the development of equitable economic prospects. The gig economy, driven by digital platforms, enables people to engage in flexible employment options. For marginalized groups who might encounter obstacles to traditional work, such as those with disabilities or those living in neglected areas, this can be very helpful. By facilitating these people's access to sources of income, IR 4.0 fosters economic inclusion while also supporting social sustainability.

Technology can also improve social sustainability by expanding access to basic services. For instance, telemedicine makes healthcare accessible in rural or underprivileged areas. To improve public services like transportation, trash management, and energy efficiency, smart cities use IoT and data analytics. In the wake of IR 4.0, however, new issues with social sustainability also surface. There are worries about unemployment and underemployment due to the possibility of a considerable number of jobs being replaced by automation and AI. To solve this, it is necessary to investigate social safety nets and policies to offer financial stability to individuals impacted by technology disruptions, such as universal basic income (UBI) or job guarantees. This guarantees that economic progress benefits all societal members and lessens social inequality.

The potential for digital exclusion is another difficulty. People without internet connection or digital skills may be marginalized as more facets of life shift online. It is crucial to eliminate the digital divide through programs that offer inexpensive internet connection and instruction in digital skills. In the digital age, privacy and data security are also essential components of societal sustainability. Concerns regarding privacy violations and misuse are raised by the massive collection and analysis of personal information. To preserve individual rights and uphold confidence in the digital ecosystem, strict data protection laws and digital ethics frameworks are required.

Additionally, the development of AI and automation has brought ethical issues into the realm of decision-making. Biases in AI systems and algorithms can exacerbate already-existing societal injustices, especially when it comes to hiring and lending decisions. Social sustainability depends on ensuring that these technologies are created and applied in ways that are impartial and fair. As a result, IR 4.0 has the potential to have a significant impact on social sustainability in both favorable and unfavorable ways. Social sustainability can be improved by the fair distribution of benefits, proactive reskilling programs, and inclusive economic possibilities. However, issues like algorithmic prejudice, data privacy, employment displacement, and digital exclusion demand careful attention and regulation. In the era of IR 4.0, achieving social sustainability necessitates a comprehensive strategy where governments, corporations, civil society, and individuals work together to harness the transformational power of technology and handle its related issues. By doing this, communities can work towards a future that meets the requirements of all current and future generations while still being socially inclusive and technologically sophisticated.

The Fourth Industrial Revolution has brought about a new era of technical development and digital upheaval. Social sustainability is just one of the many sectors of society this phenomenon has broad ramifications. The incorporation of IR 4.0 technologies to advance societal equality, justice, and well-being is referred to as "Social Sustainability 4.0." Through its potential to improve the accessibility of information and knowledge, IR 4.0 can make a substantial contribution to social sustainability. People now have access to a variety of knowledge and learning resources because to the quick developments in digital technologies like the Internet of Things (IoT) and artificial intelligence (AI). By empowering marginalized communities and bridging the digital divide, this improved access may foster social inclusion and equal opportunity for all.

In addition, IR 4.0 has the potential to transform established sectors and provide new employment opportunities. Automation and robotics are two technological advancements that can streamline different industrial processes and boost production. In addition to helping businesses, this may also lead to the creation of new jobs in burgeoning industries like cybersecurity and data analytics. Societies may ensure that the advantages of IR 4.0 are fairly spread, leading in lower unemployment rates and greater economic stability, by carefully embracing these technologies and ensuring the right skills training. Additionally, IR 4.0 integration has the potential to transform healthcare and advance social welfare. Healthcare delivery could be improved by innovations like telemedicine and wearable technology, especially in rural areas. While online consultations might lessen healthcare disparities and improve accessibility, AI-powered algorithms can help spot diseases early. Additionally, the capacity of IoT devices to gather enormous volumes of data can result in more focused public health interventions and policies, ultimately improving overall population health outcomes. The difficulties and moral issues raised by IR 4.0 and its effect on societal sustainability must be acknowledged. One significant concern is that there is the potential for inequality to rise if the advantages of technology progress are not spread fairly. While

IR 4.0 could result in novel possibilities, it also runs the risk of displacing workers and exacerbating the skills gap. To make certain that no one becomes isolated, governments as well as policymakers must be proactive in putting social safety nets in place and offering retraining programs. The possible invasion of privacy and the exploitation of personal data are further ethical issues. Strong data protection rules and regulations are required because IR 4.0 heavily relies on data collecting and analysis to defend people's right to privacy. It's crucial to strike a balance between maximizing technology's promise and defending people's liberty and privacy rights.

As a result of improved information and knowledge access, new career opportunities, and a revolution in healthcare, the incorporation of IR 4.0 in society has the capacity to promote social sustainability. The difficulties and moral questions raised by this technological revolution must be addressed, though. Societies can use IR 4.0 to advance social growth, equality, and justice by assuring equitable benefit distribution, offering suitable skill training, and protecting privacy rights. It is essential for graduate students to critically assess the consequences of IR 4.0 and actively participate in influencing its sustainable implementation for the benefit of the future.

5.6 ENHANCEMENT

Project management has become a critical discipline for businesses to effectively accomplish their objectives in a world that is changing quickly. The PMBOK, created by the Project Management Institute (PMI), has long served as the normative reference for project management procedures. However, with the arrival of the Fourth Industrial Revolution (IR 4.0) and rising sustainability-related worries, it is necessary to improve PMBOK to handle the opportunities and difficulties posed by these transformative forces. We will investigate how improving the efficacy and applicability of project management techniques in the contemporary day may be accomplished by incorporating IR 4.0 and sustainability principles into the PMBOK. With the use of project management software, cloud-based communication tools, and data analytics, project management is becoming more and more digital. Real-time monitoring, better decision-making, and increased communication among project stakeholders are all made possible by these technologies. The way projects are carried out is changing as a result of automation and AI. Previously manual tasks are now automated, which lowers errors and boosts productivity. Project data can be analyzed by AI algorithms to identify dangers and provide improvement techniques. The usage of remote work has increased because of IR 4.0, which affects how project teams cooperate and communicate. Project managers must adjust to effectively manage distributed teams. Tools for data analytics and business intelligence give project managers insightful information. They can recognize trends, make data-driven decisions, and enhance project performance.

Guidance on digital project management techniques, such as the use of project management software, collaboration tools, and data analytics, can be incorporated into the PMBOK. Project managers need to know how to use these technologies efficiently. The PMBOK can emphasize agile project management approaches and offer instructions for putting into practice adaptive project management techniques that can adjust to changing project circumstances. Security in the digital age is crucial. Project managers will be more aware of the value of safeguarding project data and systems from cyber threats if PMBOK includes cybersecurity issues.

To be sustainable, project planning and execution must consider economic, social, and environmental factors, as shown in Figure 5.6. To accomplish project goals, sustainable project management seeks to minimize negative environmental effects, advance social equality, and guarantee economic viability (Hamdan, 2022).

Projects must reduce their ecological footprints to address climate change, resource scarcity, and environmental degradation. The use of sustainable project management can lessen resource consumption and lessen negative effects on the environment. Projects may have major societal effects that are both favorable and unfavorable. Projects that are managed sustainably guarantee that human rights are upheld, diversity is encouraged, and inclusion is promoted. Resource efficiency in sustainable initiatives frequently results in cost savings. Environmental impact assessments and sustainability are subject to strict restrictions in several areas. To achieve project compliance, project managers must follow various rules (Iheukwumere-Esotu and Yunusa-Kaltungo, 2021). The TBL approach, which considers the economic, environmental, and social aspects of projects, can be

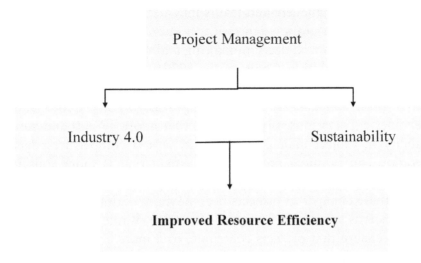

FIGURE 5.6 Enhancement via Integration of IR 4.0 and Sustainability Concept.

incorporated into the PMBOK. Through the course of the project lifecycle, this paradigm encourages project managers to maintain a balance among these three factors. As a typical procedure, environmental impact assessments may be outlined in the PMBOK. Project managers would be better able to recognize potential environmental risks and possibilities as a result (Namara et al., 2022).

The PMBOK can emphasize the value of social impact assessments, just like it does with environmental assessments. This entails assessing how projects influence local communities. Sustainable procurement practices, such as the purchase of products and services that meet moral and environmental criteria, can be a part of sustainable project management. Project managers can use PMBOK guidelines to incorporate sustainability into procurement procedures. Reporting on sustainability performance should be a part of sustainable project management. Project sustainability activities can be communicated to stakeholders using the PMBOK as a guide (Qureshi et al., 2023).

Project management procedures may become more complicated because of integrating new technologies and sustainability issues. Teams and organizations may be reluctant to accept new procedures and tools, particularly if they depart from conventional project management methods. To successfully integrate IR 4.0 and sustainability concepts, project managers and teams may need additional training and knowledge. Integration can result in improved project management, with AI-enhanced decision-making and resource optimization guided by sustainability (Chan et al., 2022). By using sustainable methods, projects might be more resistant to unforeseen social and environmental changes. By coordinating with market trends and stakeholder expectations, organizations that adopt IR 4.0 and sustainability in project management can achieve a competitive edge. Integration can promote project management that is more conscientious about the environment and society. Project managers must adjust to these changes as projects become more digital and sustainability concerns spread (El-Sayegh et al., 2020). While sustainability standards ensure that projects contribute to a more sustainable and equitable future, the inclusion of IR 4.0 can increase project efficiency. Although there are obstacles, adopting these improvements in project management practices has considerable advantages that correlate with stakeholders' shifting requirements and expectations in a world that is changing quickly (Abbaspour et al., 2018).

5.7. DISCUSSION

A fundamental framework that directs professionals in successfully managing projects across numerous industries is the Project Management Body of Knowledge (PMBOK). In recent years, there has been an increasing awareness of the necessity of incorporating sustainability and Industry 4.0 (IR 4.0) principles into project management practices. By making use of cutting-edge technologies and tackling the urgent sustainability challenges, this integration improves project outcomes. The PMBOK has historically used data-driven decision-making. Project managers can access real-time data with IR 4.0 technology, allowing for more precise monitoring and management of project development. This lowers risks and improves project predictability.

Collaboration amongst project stakeholders is made possible by the connectivity and communication capabilities of IR 4.0 tools, regardless of the participants' physical locations. Project managers can anticipate problems and take preventative action with the use of predictive analytics made possible by IR 4.0 technologies. This lowers the likelihood of expensive project delays and overruns. Robotics and artificial intelligence (AI) can automate repetitive jobs, freeing up human resources for more difficult project management responsibilities, eventually increasing project productivity. By utilizing cutting-edge sensors and monitoring systems, IR 4.0 integration can improve quality control and guarantee that project outputs meet or surpass quality standards. By considering elements like energy efficiency, resource conservation, and waste reduction, projects can be created to have as little of an impact on the environment as possible. This supports international initiatives to mitigate climate change.

Projects are urged by sustainability principles to think about how their actions may affect society. This contains elements that promote beneficial social effects, such as involvement in the community, moral labor practices, and inclusivity. Projects that are sustainable frequently have longer-term financial advantages. Project managers can make choices that eventually result in more stable financial situations by considering the costs and benefits of the life cycle. Engagement of stakeholders, especially those outside the conventional project team, is emphasized by sustainability. This encourages better connections and may result in larger project support. Regulations pertaining to sustainability are becoming more stringent in several regions. By including these factors in the PMBOK, projects are guaranteed to remain compliant with changing regulations and requirements.

The inclusion of Industry 4.0 and sustainability concepts in the PMBOK offers a thorough method for managing projects in the modern era. Project managers are given cutting-edge tools and techniques to help them negotiate the difficulties of the digital era while also making a positive impact on a more sustainable and responsible future. In a world that is evolving quickly and where technology and sustainability are key to project success, the PMBOK must be updated.

All in all, to promote innovation, efficiency, and competitiveness in modern project management practices, the Project Management Body of Knowledge (PMBOK) must be improved to incorporate Industry 4.0 (IR 4.0) and sustainability. Project managers must modify their strategies considering the global use of IR 4.0 advancements like artificial intelligence, robotics, and the Internet of Things (IoT). Professionals can take advantage of the potential of automation, real-time data analytics, and interconnectivity to optimize the utilization of resources, improve communication channels, and speed up project execution by implementing IR 4.0 principles into PMBOK. Incorporating sustainability principles into PMBOK also strengthens moral behavior to minimize negative environmental effects while providing stakeholders with long-term value. In general, this transition not only provides project managers with the instruments they need to traverse the complexity of contemporary projects but also encourages ethical corporate conduct consistent with the sustainable development objectives set forth by international authorities.

REFERENCES

Aalavi, S. & Janatyan, N. 2020. Identifying and prioritizing activities of green project management based on lean-sustainable principles in Isfahan Parks and Green Space organization. *Production and Operations Management*, 11, 1–25.

Aarseth, W., Ahola, T., Aaltonen, K., Økland, A. & Andersen, B. 2017. Project sustainability strategies: A systematic literature review. *International Journal of Project Management*, 35, 1071–1083.

Abbaspour, M., Toutounchian, S., Dana, T., Abedi, Z. & Toutounchian, S. 2018. Environmental parametric cost model in oil and gas epc contracts. *Sustainability*, 10, 195.

Adriana, T.-T. & Ioana-Maria, D. 2013. Project success by integrating sustainability in project management. In *Sustainability Integration for Effective Project Management*. IGI Global.

Armenia, S., Dangelico, R. M., Nonino, F. & Pompei, A. 2019. Sustainable project management: A conceptualization-oriented review and a framework proposal for future studies. *Sustainability*, 11, 2664.

Banihashemi, S., Hosseini, M. R., Golizadeh, H. & Sankaran, S. 2017. Critical success factors (Csfs) for integration of sustainability into construction project management practices in developing countries. *International Journal of Project Management*, 35, 1103–1119.

Buhulaiga, E., Telukdarie, A. & Albukhari, A. 2019. Success factor of multinational business transformation change management. *International Association for Management of Technology*, 1–13.

Carvalho, M. M. & Rabechini Jr, R. 2017. Can project sustainability management impact project success? An empirical study applying a contingent approach. *International Journal of Project Management*, 35, 1120–1132.

Chan, M., Jin, H., Van Kan, D. & Vrcelj, Z. 2022. Developing an innovative assessment framework for sustainable infrastructure development. *Journal of Cleaner Production*, 368, 133185.

Chawla, V., Chanda, A., Angra, S. & Chawla, G. 2018. The sustainable project management: A review and future possibilities. *Journal of Project Management*, 3, 157–170.

Coşkun, C. 2019. *Development of a Risk Assessment Method for Sustainable Construction of Megaprojects*. Middle East Technical University.

Dreyer, M., Von Heimburg, J., Goldberg, A. & Schofield, M. 2020. Designing responsible innovation ecosystems for the mobilisation of resources from business and finance to accelerate the implementation of sustainability. A view from industry. *Journal of Sustainability Research*, 4.

El-Sayegh, S., Romdhane, L. & Manjikian, S. 2020. A critical review of 3D printing in construction: Benefits, challenges, and risks. *Archives of Civil and Mechanical Engineering*, 20, 1–25.

Fernández, M. E. A. 2019. *Business Model for Sustainable Innovation at Project Level*. Universidad Del País Vasco-Euskal Herriko Unibertsitatea.

García-Gómez, F. J., Rosales-Prieto, V. F., Sánchez-Lite, A., Fuentes-Bargues, J. L. & González-Gaya, C. 2021. An approach to sustainability risk assessment in industrial assets. *Sustainability*, 13, 6538.

Ghorbani, M. K., Hamidifar, H., Skoulikaris, C. & Nones, M. 2022. Concept-based integration of project management and strategic management of rubber dam projects using the SWOT–AHP method. *Sustainability*, 14, 2541.

Hamdan, H. A. M. 2022. *Toward a Neighborhood-Level Procurement Strategy: An Exploratory Study of Project Procurement and Collaboration Practices in Sustainable and Zero-Emission Neighborhood Projects*. Doctoral thesis. NTNU Open.

Hashemi-Tabatabaei, M., Sharifpour, H., Ghaseminezhad, Y. & Amiri, M. 2022. Investigating cause-and-effect relationships between supply chain 4.0 technologies. *Engineering Management in Production and Services*, 14.

Herrera-Reyes, A. T., Carmenado, I. D. L. R. & Martínez-Almela, J. 2018. Project-based governance framework for an agri-food cooperative. *Sustainability*, 10, 1881.

Hu, W., Lim, K. Y. H. & Cai, Y. 2022. Digital twin and Industry 4.0 enablers in building and construction: A survey. *Buildings*, 12, 2004.

Ibrahim, F. S. B., Esa, M. B. & Rahman, R. A. 2021. The adoption of IoT in the Malaysian construction industry: Towards construction 4.0. *International Journal of Sustainable Construction Engineering and Technology*, 12, 56–67.

Iheukwumere-Esotu, L. O. & Yunusa-Kaltungo, A. 2021. Knowledge management and experience transfer in major maintenance activities: A practitioner's perspective. *Sustainability*, 14, 52.

Khurshid, K., Danish, A., Salim, M. U., Bayram, M., Ozbakkaloglu, T. & Mosaberpanah, M. A. 2023. An in-depth survey demystifying the

Internet of Things (IoT) in the construction industry: Unfolding new dimensions. *Sustainability*, 15, 1275.

Lance, B. 2022. Enhancing public service delivery in a VUCA environment in South Africa: A literature review. *Вестник Российского Университета Дружбы Народов. Серия: Государстве-нное И Муниципальное Управление*, 9, 418–437.

Larsson, J. & Larsson, L. 2020. Integration, application and importance of collaboration in sustainable project management. *Sustainability*, 12, 585.

Leoto, R. F. 2020. *Limits and Opportunities of Integrated Design in Sustainable Buildings: The Need for a More Comprehensive Project Process.* https://papyrus.bib.umontreal.ca/xmlui/handle/1866/23386

Marcelino-Sádaba, S., González-Jaen, L. F. & Pérez-Ezcurdia, A. 2015. Using project management as a way to sustainability. From a comprehensive review to a framework definition. *Journal of Cleaner Production*, 99, 1–16.

Marnewick, C. 2017. Information system project's sustainability capability levels. *International Journal of Project Management*, 35, 1151–1166.

Martens, M. L. & Carvalho, M. M. 2017. Key factors of sustainability in project management context: A survey exploring the project managers' perspective. *International Journal of Project Management*, 35, 1084–1102.

Martínez-Perales, S., Ortiz-Marcos, I., Juan Ruiz, J. & Lázaro, F. J. 2018. Using certification as a tool to develop sustainability in project management. *Sustainability*, 10, 1408.

Mortaheb, M. M. & Mahpour, A. 2016. Integrated construction waste management, a holistic approach. *Scientia Iranica*, 23, 2044–2056.

Namara, I., Hartono, D. M., Latief, Y. & Moersidik, S. S. 2022. Policy development of river water quality governance toward land use dynamics through a risk management approach. *Journal of Ecological Engineering*, 23, 25–33.

Obradović, V., Todorović, M. & Bushuyev, S. 2019. Sustainability and agility in project management: Contradictory or complementary? *Advances in Intelligent Systems and Computing III: Selected Papers from the International Conference on Computer Science and Information Technologies, CSIT 2018, September 11–14, Lviv, Ukraine.* Springer, 522–532.

Ohueri, C. C., Habil, H. & Liew, S. C. 2023. The current strategies for effective communication in the Malaysian construction industry. *Journal of Language and Communication*, 10, 113–128.

Piyathanavong, V., Huynh, V.-N., Karnjana, J. & Olapiriyakul, S. 2022. Role of project management on sustainable supply chain development through Industry 4.0 technologies and circular economy during the Covid-19 Pandemic: A multiple case study of Thai metals industry. *Operations Management Research*, 1–25.

Poveda-Orjuela, P. P., García-Díaz, J. C., Pulido-Rojano, A. & Cañón-Zabala, G. 2019. Iso 50001: 2018 and its application in a comprehensive management system with an energy-performance focus. *Energies*, 12, 4700.

Qureshi, A. H., Alaloul, W. S., Wing, W. K., Saad, S., Ammad, S. & Altaf, M. 2023. Characteristics-based framework of effective automated monitoring parameters in construction projects. *Arabian Journal For Science And Engineering*, 48, 4731–4749.

Qureshi, A. H., Alaloul, W. S., Wing, W. K., Saad, S., Ammad, S. & Musarat, M. A. 2022. Factors impacting the implementation process of automated construction progress monitoring. *Ain Shams Engineering Journal*, 13, 101808.

Rezahoseini, A., Noori, S., Ghannadpour, S. & Bodaghi, M. 2019. Investigating the effects of building information modeling capabilities on knowledge management areas in the construction industry. *Journal of Project Management*, 4, 1–18.

Rezghdeh, K. & Shokouhyar, S. 2020. A six-dimensional model for supply chain sustainability risk analysis in telecommunication networks: A case study. *Modern Supply Chain Research and Applications*, 2, 211–246.

Silvius, A. G. 2017. Sustainability as a new school of thought in project management. *Journal of Cleaner Production*, 166, 1479–1493.

Silvius, A. G., Kampinga, M., Paniagua, S. & Mooi, H. 2017. Considering sustainability in project management decision making; An investigation using Q-methodology. *International Journal of Project Management*, 35, 1133–1150.

Silvius, A. G. & Schipper, R. P. 2014. Sustainability in project management: A literature review and impact analysis. *Social Business*, 4, 63–96.

Silvius, A. G. & Schipper, R. P. 2016. Exploring the relationship between sustainability and project success-conceptual model and expected relationships. *International Journal of Information Systems and Project Management*, 4, 5–22.

Silvius, A. G. & Schipper, R. P. 2020. Exploring variety in factors that stimulate project managers to address sustainability issues. *International Journal of Project Management*, 38, 353–367.

Silvius, A. G., Schipper, R. P. & Planko, J. 2012. *Sustainability in Project Management*. Gower Publishing, Ltd.

Stanitsas, M., Kirytopoulos, K. & Leopoulos, V. 2021. Integrating sustainability indicators into project management: The case of construction industry. *Journal of Cleaner Production*, 279, 123774.

Sudhir, C. & Rc, G. 2023. Development and validation of TPM implementation practices in industries: Investigation from Indian SMES. *Operations Management Research*, 1–16.

Unterhofer, M., Rauch, E. & Matt, D. T. 2021. Hospital 4.0 roadmap: An agile implementation guideline for hospital manager. *International Journal of Agile Systems and Management*, 14, 635–656.

Vásquez, J., Aguirre, S., Puertas, E., Bruno, G., Priarone, P. C. & Settineri, L. 2021. A sustainability maturity model for micro, small and medium-sized enterprises (MSMES) based on a data analytics evaluation approach. *Journal of Cleaner Production*, 311, 127692.

Walters, L. E. M., Scott, R. E. & Mars, M. 2018. A teledermatology scale-up framework and roadmap for sustainable scaling: Evidence-based development. *Journal of Medical Internet Research*, 20, E224.

Yazici, H. J. 2020. An exploratory analysis of the project management and corporate sustainability capabilities for organizational success. *International Journal of Managing Projects in Business*, 13, 793–817.

Yu, M., Zhu, F., Yang, X., Wang, L. & Sun, X. 2018. Integrating sustainability into construction engineering projects: Perspective of sustainable project planning. *Sustainability*, 10, 784.

CHAPTER 6

Project Management and Sustainable Buildings in the Era of IR 4.0

6.1 SUSTAINABLE BUILDINGS

Environmental effects from the building and construction industry are substantial. It is responsible for 40% of all CO_2 emissions associated with energy. It is a sector with greater emissions, from the acquisition and manufacture of building materials to the actual construction operations. These emissions are made worse by unsustainable building techniques, which intensify the sector's environmental effects. In terms of energy usage, buildings are also important, and a considerable amount of the world's energy consumption is used for heating, cooling, lighting, and the functioning of structures. Consequently, the sector's net environmental effect is influenced by the carbon emissions produced by the energy sources utilized to power structures. Realizing the importance of sustainable construction practices and their function in reducing the detrimental impacts of climate change requires a close examination of these factors. Thus, sustainable design approaches incorporate a variety of eco-friendly elements that place a high value on resource efficiency and mindful design. Sustainable buildings encompass a holistic strategy

DOI: 10.1201/9781032621760-6

that lessens environmental effects while boosting occupant wellness, from efficient use of energy and water preservation to waste minimization and new sustainable design approaches.

A sustainable building, also known as a "green building," is the result of a design philosophy that emphasizes improving resource use effectiveness, including the consumption of energy, water, and materials, while minimizing the effects of buildings on human health and the environment throughout their lifetimes. Although the term "green building" is used in numerous contexts, it is generally agreed that it refers to structures that are planned and maintained to minimize the overall negative effects of the built environment on human well-being and the environment. This is done by making effective utilization of energy, water, and other resources, promoting worker efficiency, preserving occupant health, and minimizing waste, contamination, and environmental degradation. Therefore, sustainable architecture revolutionizes conventional building techniques in addition to design conceptualization. A prefabricated building is one example of an innovative approach that is gaining popularity. Green building techniques can change conventional construction methods by emphasizing effectiveness, minimizing waste, and recycling. This approach reduces construction waste, energy use, and construction time dramatically, while simultaneously providing improved control of quality and increased efficiency because components are made off-site and reassembled on-site.

A significant shift is needed to create a more sustainable future, and it appears that this movement is already evident in several sectors, including architecture. We can design buildings that are not only ecologically responsible but also offer a healthy and happy living and working environment by making conserving energy, water preservation, reducing waste, and the utilization of sustainable materials a top priority. The potential for sustainable architecture is enormous. Increasingly, sustainable characteristics can be integrated into the design and construction of the environments in which we live by embracing novel design approaches and building techniques.

6.2 ROLE OF SUSTAINABLE BUILDINGS IN ACHIEVING SUSTAINABILITY

Simply put, sustainability is the ability for something to keep evolving, support itself in the future, and last forever. From a human point of view, sustainability for the earth implies that it can carry on delivering what it was intended to do, namely giving us all access to potable water, clean air, and a high standard of living forever. It cannot be sustained, and that is where we are all right now. We are eroding the system on which we as humans fully rely in our effort to achieve prosperity, progress, and success; as a result, we now pose an imminent danger to our way of existence. Our Earth is a system, and everything on it—including civilization, the environment, and the economy—is interconnected. Four guidelines must be followed if mankind is to live sustainably:

1. Minimize reliance on heavy metals and fossil fuels.

2. Minimize back synthetic chemical consumption.

3. Prevent environmental devastation.

4. Always ensure we are not interfering with people's ability to adequately satisfy their necessities globally.

We must follow these basic recommendations in all we undertake, whether at home or work. We can be sustainable in the future if we diligently adhere to these basic guidelines. We shall all live better lives, produce fewer pollutants and waste products, construct more green buildings, and do other initiatives that benefit the environment. Given the need to accomplish the long-term sustainable growth of any country, green buildings play an essential part in the construction industry. The ultimate result of green buildings is high-performance buildings with efficient resource utilization and minimizing human and environmental effects. The environment and building inhabitants can both benefit greatly from sustainable building practices. Energy-efficient buildings require less energy and emit a smaller amount of greenhouse gases, which lessens the building's total environmental effect. Our reliance on fossil fuels is lessened by structures that utilize renewable energy resources like solar or wind power.

Sustainable construction techniques also have financial advantages. Buildings that use less energy can reduce their energy expenses, which can result in lower utility prices for building occupants. The sale of surplus electricity back to the grid is another way that buildings that employ renewable energy sources might make money. Sustainable construction techniques also help society. For building occupants, a healthier

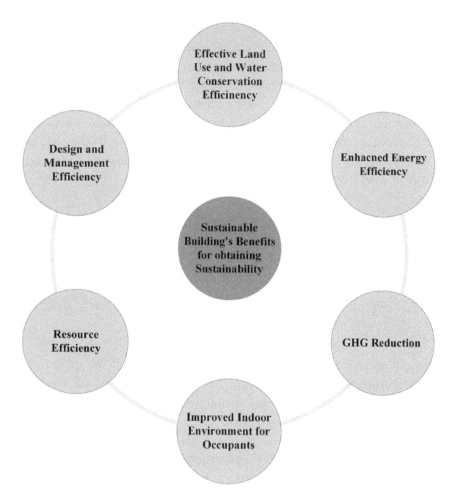

FIGURE 6.1   Role of Sustainable Buildings in Achieving Sustainability.

and cozier living environment can be provided by structures that are intended to be efficient in terms of energy consumption while employing sustainable materials. Buildings with green areas and other sustainable elements can also enhance the quality of life and encourage well-being. Moreover, numerous benefits can be obtained from sustainable buildings for accomplishing sustainability, as summarized in Figure 6.1.

6.2.1 Enhanced Energy Efficiency

To maximize energy efficiency, reduce environmental impact, and give occupants a safe, healthy, and effective place, sustainable buildings have been planned and constructed. Incorporating passive solar design elements to maximize natural light and heat, using energy-efficient lighting and appliances, installing insulation and weather stripping to minimize heating and cooling damages, and using energy from renewable sources like solar or wind turbines are just a few ways that sustainable buildings can improve their energy efficiency. As shown in Figure 6.2, the energy efficiency of buildings can significantly be improved with the incorporation of certain energy strategies. The energy efficiency strategy framework can be divided into four sections such as efficient design strategies, proactive and reactive energy management mechanisms, regulations and policies for effective application of energy efficiency initiatives, and incorporation of smart technologies. Each of these sections can be attained if the mentioned strategies are incorporated in respective sections.

6.2.2 GHG Reductions

In the construction sector, green building, also known as sustainable construction, aims to reduce the unfavorable environmental effects of building structures. Reducing carbon dioxide (CO_2) emissions from the manufacturing and consumption of materials, energy, and other resources is the main objective of green building. Green building techniques can dramatically lower CO_2 emissions by emphasizing resource efficiency and reducing possible global warming. Reducing the number of new materials utilized for construction is the most efficient strategy to lessen the adverse environmental effects of building materials. This can be accomplished through incorporating reclaimed materials in new construction projects and choosing building materials with a greater percentage of recycled material, such as recycled steel and concrete. Reusing already-existing materials is an effective strategy to lower resource usage and CO_2 levels.

6.2.2.1 Zero Carbon Emission Buildings

Carbon emissions can be classified into two broader categories such as operational carbon emissions and embodied carbon emissions. The energy required for operating the facility in occupancy, including air conditioning, heating, hot water generating, lighting, lifts, machines, cell phone recharging, cooking, and other activities, is known as operational carbon emissions. This is typically accomplished in existing structures by consuming gas in boilers or by utilizing electrical power, whereas embodied carbon emissions are

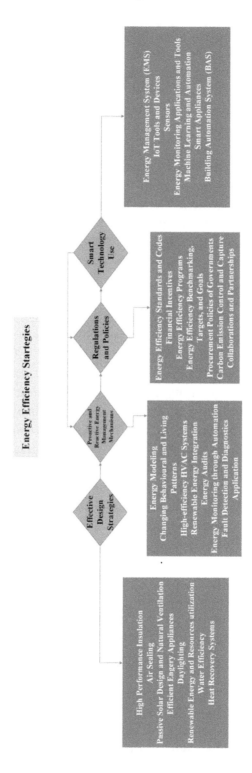

FIGURE 6.2 Energy Efficiency Strategies.

the ones produced by commodities and building materials during all phases of manufacturing, use, replacement, destruction, and disposal. The production phase is often responsible for a greater amount of carbon emission. Net-zero carbon criteria are typically attained by restricting operational energy requirements and embodied carbon emissions to the amount that is thought to be suitable for a net-zero carbon future. In this scenario, all demand can be satisfied by a combination of locally produced renewable energy and/or zero-carbon energy sources (such as power, and perhaps gas) provided via the national grid. Most of the time, it may not be feasible to satisfy a building's whole energy needs with on-site renewables, necessitating imports from the national grid.

To construct a zero-emission building, certain principles should be incorporated during the life cycle of a project. These principles include the following:

Principle of Natural Ventilation – By lowering the temperature and supplying fresh air, a well-designed natural ventilation approach can improve comfort in a building, hence minimizing cooling costs and minimizing the need for air-conditioning.

Principle of Reflective Surfaces – Although the sun is a potent source of light, it also generates a lot of heat. On a building facade, striking an appropriate equilibrium between lighting and ventilation, as well as between windows and walls, assists in maximizing daylight while reducing unwanted heat transfer.

Principle of Energy-Efficient Appliances – Utilizing lighting controls, such as sensors that detect occupancy or photoelectric sensors that only turn on lights when there is insufficient natural light, lowers energy usage.

Principle of Water Conservation Appliances – Installing low-flow faucets in sinks and energy-efficient toilets can help you save water. Water usage is reduced through tap aerators and shut-off controls without compromising operation. The consumption of water for flushing is also reduced by dual-flush toilets.

Principle of On-Site Energy Production – On-site energy production through solar panels can significantly be a source of cost efficiency.

Principle of Low-Carbon Materials–The energy in building materials contributes significantly to the overall carbon emissions of a structure. Embodied energy is the energy needed for the extraction, processing, and transportation of the raw materials used in the building's construction and maintenance. The utilization of recycled materials is another option in addition to using less energy-intensive raw materials.

Principle of Microenvironment – Utilization of nature to make the buildings sustainable such as using plants, trees, and green landscapes.

6.2.2.2 Net Zero Emission Buildings

A zero-net energy building refers to a structure that uses clean renewable resources to produce energy on-site in an amount that is equal to, or more than, the total amount of energy used on-site over a year. In other words, the goal of net zero is to "balance" or "cancel out" any carbon that we produce. When the quantity of greenhouse gas we create equals the amount that is removed, we have reached net zero. Zero carbon refers to the emissions that come from a good or service; it signifies that no carbon is emitted at all.

6.2.2.3 Carbon-Neutral Buildings

A building is considered carbon neutral if no greenhouse gases that contribute to climate change are emitted during its design, construction, or use. Meeting the government's climate goals will be considerably aided by reducing damaging greenhouse gases released by the building industry. One of the simplest methods for rendering the building carbon neutral is to use renewable energy. You can either create your energy on-site or purchase it from green, off-site vendors. Emissions will be reduced as well as your energy costs if you decide to invest in producing your renewable energy on-site. The recovery process periods for renewable energy facilities have drastically decreased because of the skyrocketing cost of electricity.

There are several different renewable energy options available for investment. One can use energy from waste materials, anaerobic digestion, wind, biomass, solar energy, hydropower, and a variety of other sources on their property.

6.2.2.4 Nearly Zero-Emission Buildings

The term "nearly zero-emission building" refers to a structure with extremely high energy efficiency, and the nearly zero or minimal amount of energy needed should be entirely or largely met by renewable energy, which includes energy generated on-site or in the vicinity.

Consequently, there is a pressing requirement for sustainable building techniques that will contribute to lowering carbon dioxide levels as the environment keeps getting worse because of the excessive buildup of carbon dioxide in the atmosphere. One of the most efficient methods to achieve this is using green building techniques, which provide several advantages, including lower energy use, better air quality, and other things.

6.2.3 Improved Environment for Occupants

A key component of designing a green building is enhancing interior air quality. You may significantly lower the concentration of carbon dioxide in indoor air by putting sustainable practices in place, like better ventilation and air filtering systems. Additionally, the likelihood of off-gassing volatile organic compounds, a significant factor in poor indoor air quality, can be decreased by choosing natural materials. Green construction techniques can benefit both the environment and the well-being of building inhabitants by enhancing indoor air quality. This can be achieved by focusing on sick building syndrome (SBS), thermal comfort, acoustic comfort, and visual comfort.

6.2.3.1 Sick Building Syndrome (SBS)

The term "sick building syndrome" (SBS) refers to a condition whereby a person's place of employment or residence causes them to become ill or infect them with a chronic disease. Buildings must be planned to minimize occupant exposure to chemicals, the indoor environment must be constantly monitored, and all water pipes and damp areas must be under constant observation and control to prevent water leaks. Inspection of water lines will result in less moisture, which is the primary contributor to a lot of problems with microbial growth and mites. For optimal indoor environmental quality (IEQ) and a decrease in SBS, it is also essential to choose appropriate third party–certified materials for the construction of buildings.

6.2.3.2 Thermal Comfort

The most significant and straightforwardly stated IEQ parameter probably involves thermal comfort. The working environment must be thermally comfortable for employees to perform to the best of their abilities. Thermal comfort, on the other hand, is dependent on everyone's ability to regulate their body temperature, which is influenced by a variety of variables, including gender, race, age, place of residence, climate, and season. People's moods are directly impacted by their sense of comfort in an interior setting. Working under ideal conditions improves our ability to concentrate and perform in workplace environments, and thermal comfort has a positive impact on both well-being and productivity. Sustaining perfect thermal comfort also involves removing any health risks, which is extremely significant.

6.2.3.3 Acoustic Comfort

When planning and designing a project, the acoustical environment of a building is sometimes given little to no consideration. Typically, the designer's main priorities are the residence or workspace's usability and aesthetics. The elements influencing the efficiency of the personnel using the building are occasionally disregarded. Individuals work at their best in a peaceful residence and employees have lower absenteeism rates when they are in a comfortable work environment. The quality of the indoor environment, the outside temperature, acoustics, and natural daylight and electric lighting are a few of the aspects that contribute to a comfortable place to live in. Noise from vehicles outside, mechanical noise from neighboring areas, phones, and conversations inside offices or homes can all cause hearing damage with acute irritation. These challenges can be encountered through effective acoustic management. The acoustic management can be divided into five basic steps, as shown in Figure 6.3.

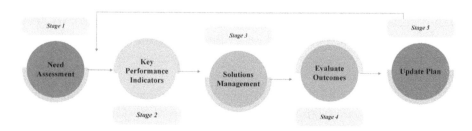

FIGURE 6.3   Acoustic Management.

It can be seen from Figure 6.3 that understanding the acoustic requirements of your building's inhabitants according to their behaviors, choices, and expectations is the first step. There may be a range of requirements for sound insulation, ambient noise, resonance, and speech intelligibility, depending on the type of environment. Evaluate the acoustic performance of your building areas in the second stage utilizing both objective and subjective techniques. The physical characteristics of sound, such as sound pressure threshold, sound transmission class, reverberation duration, and voice transmission index, are quantified using tools and software utilizing objective methodologies. In subjective approaches, the perceptual qualities of sound, such as loudness, irritation, convenience, and contentment, are assessed utilizing scales and questionnaires. To compare the real and desired acoustic performances, you can utilize both techniques.

Implementing acoustic measures that can increase building occupants' comfort with sound is the third phase. The three categories of acoustic solutions are origin, path, and receiver. The goal of solutions management is to lessen or regulate the sound sources, such as people, equipment, and machinery, that produce undesirable noise. The goal of path solutions is to stop or absorb sound waves as they move through the atmosphere or the building itself, such as walls, floors, ceilings, or windows. Receiver solutions, like headphones, speakers, or soundscapes, try to amplify or reduce the sound signals that reach the listeners' ears. The fourth phase entails assessing the acoustic results of your acoustic solutions using both objective and subjective criteria. Repeating the assessments and evaluations from the second phase will allow you to compare the outcomes between the acoustic solutions' implementation and those without them. To be able to determine whether the residents of your building are happy with the sound environment and whether they have observed any modifications in their health, well-being, productivity, or communication, you can additionally observe them and solicit feedback from them. The building's acoustic plan needs to be updated as the fifth step, depending on the assessment of the acoustic results. Acoustic comfort is an evolving process that can alter over time owing to a variety of reasons, including occupancy, consumption, maintenance, remodeling, or outside noise. Acoustic comfort is not a static situation. To make sure that it fulfills the present and future acoustic requirements of your building's inhabitants, you should frequently examine and amend your acoustic plan.

6.2.3.4 Visual Comfort

When constructing buildings that support occupant well-being, visual comfort should be considered in a significant manner, just like thermal insulation, acoustic comfort, and air quality. It's important to consider how lighting affects individuals when designing a space. Natural light greatly affects how we perceive, interpret, and respond to situations. It has been demonstrated that adequate lighting conditions—including brightness, vistas of the outside world, and the absence of glare—have a favorable effect on mood and productivity. In a room, there must be sufficient light for people to be inspired to work, creative, and, most importantly, well-focused.

6.2.4 Resource Efficiency

Efficient utilization of resources can play a critical role in sustainable development. The term "resource efficiency" refers to the connection between natural raw materials or technical-economic materials and the advantages derived from using them, whether for production or consumption. The goal is to maximize a product's or service's advantages while reducing use and waste. Therefore, the effective management of natural resources for the betterment of everybody in the community is sustainable resource use. The basic goal of sustainable development is to meet the requirements of the present generation without jeopardizing those of the future.

Without intervention from humans, natural resources would already exist. Some renewable ones can be easily regenerated, and some non-renewable ones cannot. The economics of a region is influenced by its resources. A nation can preserve these resources for future generations if these resources are used wisely. However, given the current conditions, it is extremely unlikely that future generations and emerging nations will be able to obtain their fair share due to the indiscriminate exploitation of our contemporary resources. Furthermore, the repercussions are terrible, and the impact on the ecosystem will cause serious harm that exceeds the environment's carrying capacity. Figure 6.4 highlights four crucial stages such as saving resources, recycling and reusing, substituting resources, and reducing to achieve resource efficiency by any organization globally. With the inclusion of resource efficiency principles in resource efficiency frameworks such as eco-design, eco-inventions, green procurement, and sustainable development practices, outcomes such as cost

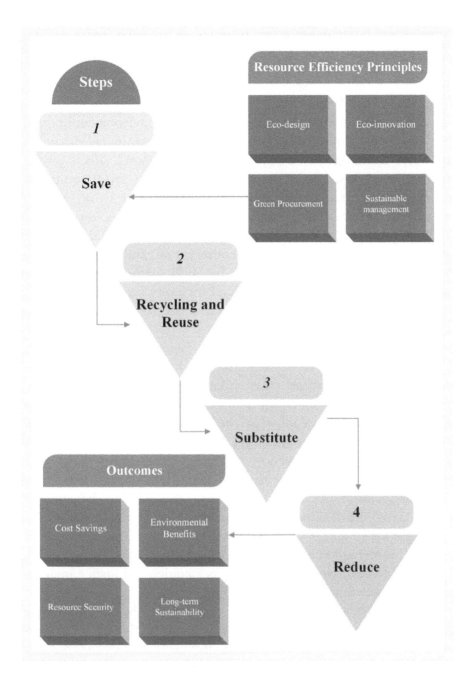

FIGURE 6.4   Resource Efficiency Framework.

savings, environmental benefits, resource security for future generations, and sustainable development can be attained.

6.2.5 Design and Management Efficiency

The challenges caused by not designing with the environment in mind can be solved through the implementation of sustainable design. It comprises planning with an eye on reducing negative effects on the environment, conserving resources, and advancing social equity. By incorporating sustainable design principles into our work, we can reduce carbon emissions, waste, and use of resources. Sustainable design strategies include aspects like utilizing renewable energy resources, making plans for energy conservation, and including green infrastructure in cities and towns. To build surroundings that are easily accessible, secure, and beneficial for every segment of society, sustainable design also takes the social impact of development into account.

Unfortunately, a lot of architects and designers still prioritize aesthetics and expense over sustainability, which harms the environment by wasting resources, raising carbon emissions, and causing environmental damage. Future consequences for our world could be severe because of changes in the climate, loss of habitat, and biodiversity decline. Furthermore, because disadvantaged groups are more likely to be harmed by unfavorable environmental consequences of growth, ignoring the environment in design frequently results in social injustice. We run the risk of escalating the environmental crisis and leaving future generations with an unstable and unsustainable planet if we carry on designing without considering the environmental consequences. Design decisions that overlook ecology have negative effects on society's equity and justice in addition to the environment.

The future sustainability of the earth depends on sustainable design. To create a resilient, effective, and productive built environment for individuals and the planet, architects and designers must place a high priority on sustainability in their work. The built environment we create now will have a lasting effect on future generations. We must take immediate action to ensure that the structures and locations we design are sustainable and avoid contributing to further environmental degradation, since we can no longer ignore the environmental impact of our designs.

6.2.6 Effective Land Use

The idea of sustainable land management is to acknowledge that land is a limited resource and that using it shouldn't jeopardize other generations' potential to benefit from it. Utilizing land resources is intended to promote social progress, environmental sustainability, and economic development. This calls for a comprehensive strategy that considers the interactions between various land applications, such as agriculture, forestry, and urban growth, as well as the effects of human activity on the environment. To be able to fulfill the demands of both the present and future generations, sustainable land management strikes a balance between social, environmental, and economic variables. It entails using techniques and technology to preserve or increase the land's capacity for production while preserving and enhancing its natural resources, such as water, land, and habitat.

Various methods can be used for effective land management to attain sustainability in the built environment. Figure 6.5 depicts

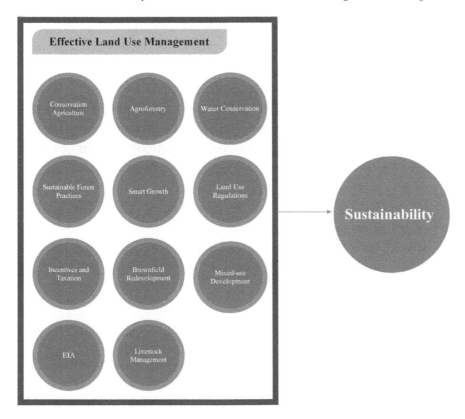

FIGURE 6.5 Land Use Management Framework.

these methods which include conservation agriculture, agroforestry, water conservation management, sustainable forest practices, smart growth, land use regulations, incentives and taxation system, brownfield redevelopment, mixed-use development, environmental impact assessments (EIA), and sustainable livestock management.

6.2.6.1 Conservation Agriculture

Through no-till or reduced tillage techniques, conservation agriculture entails minimizing soil disturbance, enhancing soil cover through crop residues and mulching crops, and rotational crops. This promotes retention of water, decreases erosion, and preserves the structure and health of the soil.

6.2.6.2 Agroforestry

By intercropping or planting trees on the boundaries of or between fields, agroforestry refers to the integration of trees into agricultural landscapes. Increased soil organic matter, windbreaks, shade, and additional income from fruit and timber production are all benefits of trees.

6.2.6.3 Water Conservation Management

Conserving water resources, lowering contamination of water, and increasing irrigation effectiveness are all aspects of sustainable water management. Rainwater collection, drip irrigation, and contour agriculture can all help with this.

6.2.6.4 Sustainable Forest Practices

When trying to manage forests in a way that balances economic, social, and environmental goals, sustainable forest practices must be used. Maintaining the health and productivity of forests, lowering carbon emissions, and fostering biodiversity are all possible with sustainable forest management.

6.2.6.5 Smart Growth

Smart growth involves building communities and societies with accessible transportation infrastructure, green space management, and the provision of equal opportunities to the community. These strategies will assist in the attainment of effective land use management.

6.2.6.6 Land Use Regulations
With the employment of standards, codes, and building regulations, zoning can be done in urban and rural areas. This will help in avoiding population density issues and land development practices.

6.2.6.7 Incentives and Taxation System
Sustainability can be achieved through effective land use management with the implementation of incentives and recognition systems from governments and global institutions. Moreover, penalties and fines should also be imposed on violations to encourage the developers to use the land efficiently. Furthermore, enhanced taxation mechanisms directly targeting the land use management strategies should be introduced and implemented.

6.2.6.8 Brownfield Redevelopment
Brownfields are the abandoned or contaminated sites polluted by the industries. Therefore, the reclamation and redevelopment of such sites can play a very important role in making the built environment a sustainable one.

6.2.6.9 Mixed-Use Development
When two or more functions are combined in one structure or broad area, this is known as a mixed-use development. The phrase is frequently used to describe construction which integrates residential with commercial or even industrial usage, but it can also apply to amenities for the public as well as cultural and institutional functions. Mixed-use developments are frequently described as walkable and pedestrian-friendly, giving locals additional opportunities for housing, employment, and shopping in the same area and reducing their reliance on driving.

6.2.6.10 Environmental Impact Assessments (EIA)
Before a decision is made to proceed with the recommended action, an environmental impact assessment (EIA) evaluates the environmental effects of a plan, policy, program, or real project. The findings of an EIA report can be beneficial for effective land use management.

6.2.6.11 Sustainable Livestock Management
Agriculture for livestock is evolving quickly. The need for protein is rising due to continued population expansion. The ramifications of

meeting this need, though, are what should most worry us. Businesses involved in livestock are under pressure from modern consumers' concerns for sustainability. The ability to demonstrate that farming practices don't hurt the environment is one of the primary difficulties facing industry today. Consumers are also interested in the living conditions and welfare of animals. When it comes to choosing food products, people place a higher value on animal welfare than other factors. According to research, customers believe that animal-friendly products are wholesome, delicious, clean, secure, acceptable, legitimate, and more.

Businesses that deal with livestock must exhibit sustainability in their operations to attract customers because of the evolved demands of customers in the context of sustainability. Adopting contemporary methods will help you achieve this. The livestock industry may become more transparent, effective, and ecologically friendly with the use of smart technology alternatives.

6.2.7 Water Conservation Efficiency

Water efficiency is the practice of minimizing water usage by calculating the volume of water needed for a specific task and comparing it to the volume of water that is being utilized. In contrast to water conservation, water efficiency emphasizes minimizing waste rather than limiting consumption. When it comes to water efficiency, solutions should not only aim to use less potable water but also to consume less non-potable water when necessary (for example, when flushing toilets or watering plants). It also highlights the impact that customers may have on water efficiency through simple behavioral adjustments to prevent water wastage and product selections that use less water.

Water, being the most important resource for our survival, must be conserved through proficient conservation efficacy efforts. Having said that, Figure 6.6 illustrates important water conservation strategies for attaining sustainability in the built environment. The key components of the water conservation framework comprise regulations and policies, R & D initiatives, adoption of smart technologies, facilitating organizations and agencies for successful implementation of water conservation, and bringing improvement through education and awareness workshops and programs.

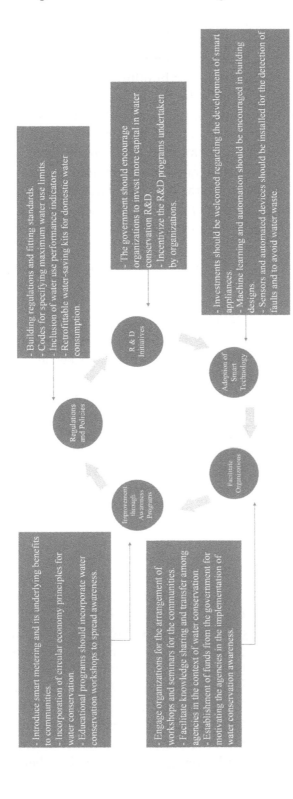

FIGURE 6.6 Water Conservation Framework.

6.3 SUSTAINABLE BUILDING MANAGEMENT

One global trend that has drawn a lot of attention recently, especially from those with an interest in the building industry, is the tendency towards sustainability and a better living for future generations. Urban and industrial growth combined with increased resource use harms the planet's capacity to replenish its resources, putting future life in peril. Recent global warming–related calamities including forest fires, floods, and torrential rains have increased interest in sustainability as a strategy to combat the problem. The search for alternative energy sources and the development of theories that seek to lessen dependency on oil energy and rationalize coal and gas consumption for electricity production has been sparked by the prohibitive price of fossil fuel energy and its detrimental effects on the environment. In addition to minimizing environmental effects, sustainable building management aims to create environments that are pleasant to be in and encourage productivity. Innovative approaches and techniques that can maximize efficacy, effectiveness, and resilience are increasingly necessary as the market for green buildings becomes higher.

The demand for a comprehensive system for managing all the project's resources, stakeholders, records, finances, and requirements has increased along with the increasingly complex nature of projects in general and construction projects in particular. Sustainable buildings and project management go hand in hand. The exclusion of any one of these essential elements will result in the non-attainment of sustainable development goals. The following are different strategies that can help in the development of sustainable buildings, but their successful execution relies heavily on effective project management:

1. **Integrated Design Approach**

   The technique that enables you to handle projects in an integrated and collaborative approach to achieve an ideal performance is known as an integrated design approach. A broad group of design experts are brought together under integrated design to work on a single project. This method helps projects since a wider range of professionals participate as a team throughout the project rather than working alone. To be able to avoid the likelihood of costly conflicts once design development or implementation gets underway, early integration is essential. Every stakeholder, especially design professionals, clients/owners, regulatory authorities, and others, is involved in the integrated design method. Construction managers, contractors, and cost analysis experts could also be involved. Whatever the project type, an integrated strategy ensures a comprehensive result rather than the accumulation of interdependent components.

EXAMPLE 6.1 THOUGHT BOX FOR STUDENTS

A transportation engineer or planner oversees a project involving a road network. A variety of government agency officials, engineers, environmental scientists, right-of-way experts, and urban designers/landscape planners may be included on the project team.

---

2. Adoption of Circular Economy Principles

Throughout their life span, circular buildings are constructed to make the most efficient use of resources and have minimal adverse effects on the environment. Materials that are long-lasting, renewable, recyclable, or biodegradable that have minimal embodied energy and carbon should be chosen to accomplish this. Components that are prefabricated, standardized, or modular can be put together, taken apart, or changed with ease. Dependence on fossil fuels and electricity from the grid can be decreased by using energy from renewable sources, such as wind turbines, solar power plants, or geothermal heat systems.

EXAMPLE 6.2 THOUGHT BOX FOR STUDENTS

Many organizations globally have adopted the circular economy principles in their day-to-day operations for improving their companies' brand name and productivity. Following are some of the organizations that are implementing circular economy:

IKEA Sweden – They recycle the old products to make new products and sell them to the clients.

The Renault Group – They utilize recycled and remanufactured car parts and design their cars with quick reassembly or disassembly.

Tesla – They use recycled products and manufacture environment-friendly cars.

Coca-Cola – They use sustainable packaging and focus on using recycled plastic for making bottles.

3. Adoption of Renewable Energy Resources

As the world struggles to mitigate the harmful effects of climate change and lessen its dependence on exhaustible and polluting fossil fuels, renewable energy solutions have gained importance. To promote sustainable development, which attempts to meet the demands of the present generation without jeopardizing the future, it has been recognized that the utilization of renewable energy sources such as solar energy, wind energy, geothermal energy, and biofuels is a critical component. By lowering greenhouse gas emissions, enhancing energy security, and giving historically energy-deprived areas access to energy, renewable energy innovations play a critical role in sustainable development. Additionally, these technologies foster economic growth and employment creation, both of which are crucial for sustainable development. Other key benefits of the adoption of renewable energy resources include the following:

- Betterment in energy security
- Provision of access to renewable energy to the communities
- Provision of new employment opportunities
- Efficient development of rural areas
- Decentralization of the production of energy
- Technological diversity

4. Indoor Environment Quality (IEQ)

The conditions inside the structure are most simply referred to as the indoor environmental quality (IEQ). In addition to access to daylight and vistas, pleasant acoustics, occupant management of lighting, and thermal comfort are also included in good air quality. It may also involve the space's functionality, such as if the design makes it simple to reach people and tools whenever required and whether there is enough room for occupants. Building executives and operators can improve building occupant contentment by considering all IEQ factors rather than just concentrating on temperature or air quality.

5. Consistent Feedback and Improvement Mechanism

Sustainable building management can be enhanced with the inclusion of consistent feedback from the customers. For smooth application, an efficient feedback mechanism must be in place by the concerned organizations. The feedback process is a never-ending tool that can guide the stakeholders in improving their processes to achieve a sustainable built environment.

6. Biophilic Design Implementation

Through the integration of direct nature, indirect nature, and space and place conditions, biophilic design is a concept employed in the construction industry to promote occupant connectivity to the natural environment. This concept is asserted to have minimal downsides when applied at both the city and building scales and to have positive effects on people's health, the environment, and the economy. The goal of biophilic design is to promote a person's affinity for the natural world that encompasses them intending to develop structures and environments that enable peaceful, naturally delightful experiences for those who utilize them. The discipline has several advantages for both physical and mental health, as well as a favorable effect on the environment because of improved sustainability efficiency. Kellert (2008) proposed six fundamental components of a biophilic design, as shown in Figure 6.7.

- Environmental Aspects

   Biophilic designs should mimic the environmental features that should relate to the natural built environment.

- Natural Shapes and Forms

   The building design should incorporate and represent the natural characteristics of both internal and external designs.

- Natural Patterns and Processes

   This element incorporates the properties of nature to enhance the man-made built environment.

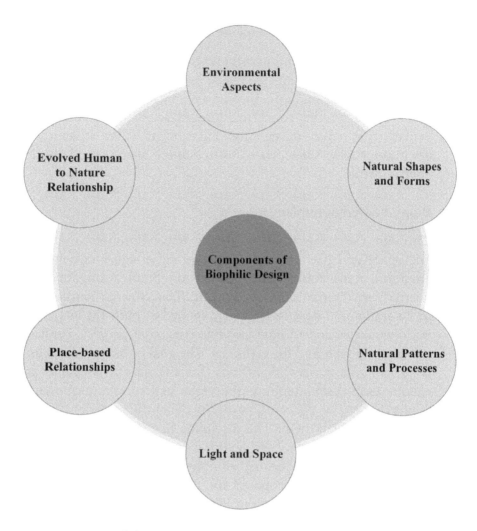

FIGURE 6.7   Biophilic Design Components.

- Light and Shape

  There are numerous ways to utilize light and space in a space, which are essential components of biophilic design.

- Place-Based Relationship

  This element of biophilic design incorporates the culture of people with ecology because people are emotionally attached to the places.

• Evolved Human-to-Nature Relationship

This element focuses on aspects of human relationships with the environment and how they can be reflected in built environments.

EXAMPLE 6.3 THOUGHT BOX FOR STUDENTS

Following are different projects that incorporated biophilic design concepts:

1. Amazon Spheres, Seattle, located in USA.
2. Bosco Verticale, Milan, located in Italy.
3. Singapore's Gardens by the Bay, located in Singapore.
4. The Edge, Amsterdam, located in Netherlands.
5. Sino-Ocean Taikoo Li Chengdu, located in China.

6.4 SUSTAINABLE BUILDINGS IN THE ERA OF IR 4.0 FOR ENHANCING THE PROJECT MANAGEMENT

Buildings that are safe for the environment and resource-efficient in their design, construction, and operation are sustainable. They combine energy-efficient technologies, utilize sustainable building materials, and reduce the production of waste and pollutants. Sustainable buildings, in the context of Industry 4.0, make use of cutting-edge technologies like the Internet of Things (IoT), artificial intelligence (AI), and data analytics to optimize the consumption of energy, optimize occupant comfort, and boost the sustainability of buildings. The incorporation of IR 4.0 in the buildings can transform them into sustainable-oriented buildings. This will have a considerable impact on revolutionizing the project management practices. Furthermore, the reduction of pollutants and carbon emissions is simply one aspect of sustainability, even though they frequently receive the most emphasis. Developing and growing in a way that is both environmentally friendly and socially appropriate is what sustainability is all about. Construction and manufacturing may now genuinely embrace a sustainable future because of smart technology from the fourth industrial revolution, or Industry 4.0.

According to the Sustainable Development Goals of the United Nations, sustainability is the fundamental business approach for the future. They involve supporting initiatives like intelligent manufacturing, environmentally friendly construction, and minimally disruptive industrialization. Industry 4.0 addresses this by fusing digital technology, artificial intelligence (AI), and the Internet of Things (IoT) with traditional operations and manufacturing. However, businesses and organizations must act now before the environment deteriorates, if they want to ensure a sustainable future. Even if innovations like cloud computing, IoT, and machine learning are unquestionably important components of the equation, the term "Industry 4.0" doesn't always relate to a particular category of technologies. These innovations highlight Industry 4.0's main objective, the fusion of physical processes and digital communication. For instance, networked IoT sensors in a sustainable and technologically advanced structure could self-regulate the consumption of energy depending on current demand and utilization.

Industry 4.0 is also characterized by data and data analytics. Customers, devices, and interconnected gadgets all continuously generate data. Organizations can further improve efficiency, optimize operations, and foster growth due to big data technologies like AI, machine learning, and data processing in real time. This can result in operations of industries running more efficiently, using less energy, and emitting less carbon as a result—all factors that contribute to sustainable development. Smart vehicles can transfer data to a cloud server for assessment throughout the supply chain, which is another area where Industry 4.0 in manufacturing is prominent. The most environmentally friendly routes or transport options can then be determined by an AI system. Environmental controls in companies under Industry 4.0 go much beyond simple heating and cooling. Intelligent filtration sensors may be able to recognize an increase in a particular chemical or pollutant and immediately take action to safeguard the health and safety of industry workers.

Moreover, planning and design, monitoring, and continuous supervision are among the areas in which Industry 4.0 is used in the building industry. Building information modeling (BIM) software is one of the technologies that architects currently employ to optimize structures for sustainability. They can incorporate organically repairing features like eco-friendly microorganisms into the design of commercial and industrial buildings. With the help of Industry

4.0 technologies, sustainable design and architecture are now possible. Upon completion of a building, heating and cooling devices are capable of real-time self-regulation for energy efficiency. Sensors and IoT devices will alert the environmental control networks to switch off the lights and cooling systems whenever people leave a room. Industry 4.0 is revolutionizing sustainable construction since smart buildings can function sustainably. Buildings may eventually be completely closed loop in terms of resource utilization, consumption of energy, and sustainability because of Industry 4.0 technology. The effluent may be recycled and purified automatically. Additionally, artificial intelligence (AI) algorithms will consistently analyze data to implement modifications in real time for more environmentally friendly utilization of energy.

Consequently, business expansion and environmental sustainability must no longer be antagonistic with the advent of Industry 4.0. Construction and manufacturing are, in many respects, the pillars of the world's economy. To support those industries' robust, intelligent, and sustainable growth, business executives simply must turn to technologies like integrated factory machinery, AI-enabled energy sources, and industrial IoT devices. To sum up the earlier discussion, we have proposed a framework in Figure 6.8 in which the attainment of a sustainable building with the integration of IR 4.0 has been illustrated. Through the incorporation of IR 4.0 tools and methods, the probability of achieving highly sustainable structures can be ensured with key characteristics such as energy and resource efficiency, automation efficiency, manufacturing and supply chain efficiency, and construction efficiency. On the other hand, only IR 4.0 inclusion does not help in making structures sustainable; therefore, the incorporation of circular economy principles such as reduce, reuse, recycle, recover, redesign, and remanufacture along with resolve can make this process of achieving sustainable structures even faster. Even then, the complete sustainable buildings are hard to achieve; thus, to move even closer to this milestone, BIM 4.0 is also made part of this framework to further automate the processes. Once all three main components of this framework such as IR 4.0, circular economy, and BIM 4.0 are not incorporated in the planning of any project, the probability of the attainment of sustainable buildings can be increased significantly. These sustainable buildings have a direct impact on the achievement of sustainability.

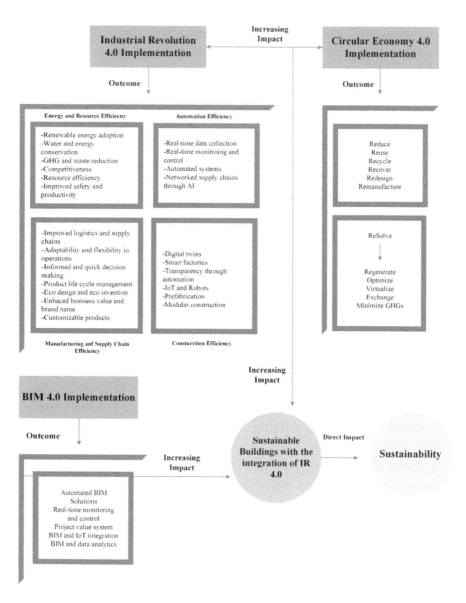

FIGURE 6.8   Sustainable Buildings in the Era of IR 4.0.

6.5 CASE STUDIES

This chapter has discussed in detail the major concepts regarding sustainable buildings. In this section, we will highlight some real-life projects globally as examples to help students observe which techniques have been adopted in these projects to achieve sustainability.

All these case studies have adopted circular economy principles, zero-carbon emission strategies, and biophilic design concepts.

6.5.1 Bahrain World Trade Center, Bahrain

A trailblazing instance of a sustainable skyscraper is the Bahrain World Trade Center. Between its twin towers, this landmark building has three enormous wind turbines that harvest wind energy to provide renewable electricity, reducing its dependency on conventional energy sources and helping to lower greenhouse gas emissions.

6.5.2 Museum of Tomorrow, Brazil

The Museum of Tomorrow, developed by famous architect Santiago Calatrava, symbolizes the idea of a sustainable structure that blends with its environment. The museum has interactive climate change and sustainability-oriented exhibits and displays, as well as solar energy systems installed on the roof and a rainwater collecting system.

6.5.3 Pixel Building, Australia

The Pixel Building, a high-design sustainable building with a Melbourne address, serves as an example of how environmental design concepts can be included. This structure makes use of cutting-edge sustainable technologies like solar panels, more effective insulation, and rainwater collection devices. To further emphasize its dedication to resource preservation and minimized waste, it also has a distinctive exterior built from recycled aluminum cans.

6.5.4 Bullitt Center, United States

The Bullitt Center in Seattle, Washington, is an illustration of a net-zero energy structure. The structure uses integrated renewable energy technology, such as solar panels on the roof and geothermal cooling and heating processes, and produces as much energy as it uses.

6.5.5 The Crystal, United Kingdom

The Crystal, a spectacular illustration of sustainable design, is situated in London. The structure emphasizes environmental awareness and education, acting as both a venue for exhibitions and a display of sustainable practices. It features several energy-saving technologies, including efficient HVAC systems, solar panels, and rainwater collection.

6.5.6 Sun-Moon Mansion, China

The Sun-Moon Mansion is the largest solar-powered structure in the world, and it is in Dezhou, China. The building's different activities are powered by a sizable array of solar panels on its roof, which, when paired with passive cooling and heating techniques, significantly reduces energy use.

6.5.7 Suzlon One Earth, Pune, India

Leading wind energy provider Suzlon Energy Limited is headquartered in Pune, India. They promised to build the greenest office campus in India using sustainability as their offering. Living up to the company's goal of "powering a greener tomorrow," only environmentally friendly and recycled materials were used in the design. All on-site resources—water, energy, air, sewage, and trash—are handled sustainably. Wastewater, sewage, and rubbish are all recycled on the premises; none is taken away.

6.5.8 One Angel Square, Manchester, UK

One Angel Square in Manchester is extremely special since it has a double-glazed facade and many stories of winter gardens that serve as breakout areas and significantly improve the working atmosphere. The structure has a sizable full-height atrium that offers good natural light throughout. The structure houses the first regional office block in the UK to receive a BREEAM "Outstanding" rating. It uses natural ventilation systems, rainwater harvesting, and renewable energy resources.

6.5.9 The Edge, Amsterdam, Netherlands

The building is BREEAM-certified, and it is regarded as the world's smartest and greenest structure. It includes 28,000 sensors that are linked to a network that not only manages the movement of people and goods within the building but also gathers and assesses information on social behavior. Compared to similar office buildings, the Edge consumes 70% less electricity. The largest system of photovoltaic panels of any office building in Europe is integrated into the roof and the south-facing facade and the entire energy needed for heating and cooling is stored in an aquifer thermal energy system. This storage system's efficiency was substantially improved by adding a heat pump.

6.5.10 One Central Park, Australia

One Central Park, a landmark example of sustainable architecture and urban vegetation, is situated in Sydney. A heliostat system that reflects sunshine into gloomy regions and large vertical gardens are both features of this high-rise residential structure. These characteristics not only boost the building's architectural appeal but also help to clean the air and use less energy.

REFERENCE

Kellert, S. R. 2008. Dimensions, elements, and attributes of biophilic design. In *Biophilic Design: The Theory, Science, and Practice of Bringing Buildings to Life*. John Wiley & Sons, Inc.

CHAPTER 7

Enhancement to PMBOK Knowledge Areas with IR 4.0 Perspective

7.1 URBANIZATION AND I.R 4.0

The Industrial Revolution and urbanization are two interrelated developments that have had a significant and long-lasting influence on the development of human civilization. The transition from rural to urbanized industrialized cultures was enormous, and it reshaped the landscape of society, economy, and culture throughout the late eighteenth and early nineteenth centuries because of the confluence of technological developments, economic changes, and social transformations.

The Industrial Revolution, frequently regarded as a watershed in human history, saw a shift away from manual labor and rural economies towards mechanized production and urbanized existence. The introduction of technology, the development of steam power, and mechanized textile production changed how productive certain sectors were. This innovation wave not only improved production efficiency but also resulted in the clustering of industries in metropolitan areas. As centers of economic activity, factories began to draw people from the countryside looking for work.

Urbanization, or the movement of people into urban regions, changed the demographics of societies and created the framework

 DOI: 10.1201/9781032621760-7

for contemporary cities. Urban landscapes and social dynamics underwent tremendous change because of urbanization, which was propelled by industrialization's needs. Cities grew larger and more complicated as people from the countryside relocated there in search of employment. Skyscrapers, factories, and infrastructure developed, transforming the surrounding landscape and giving cities their own unique identities. Urban regions experienced problems like overpopulation, substandard housing, and unhygienic conditions because of this rapid growth. However, they also opened fresh doors for technical advancement, cultural interaction, and social mobility.

Urbanization and the growing world population have increased the demand for industrialization to support social, economic, and environmental activities. It is now possible to produce industrial items more swiftly and adaptably with effective management because of technological advancements, which are essential for lowering environmental, social, and economic problems. The importance of project management has increased since each project in the construction industry is unique (Hessing and Summerville, 2014; Alaloul et al., 2018; Shahroom and Hussin, 2018).

The key elements of project management are known as project management knowledge areas, and they serve as the foundation for organizing projects and obtaining desired results (Ashaari et al., 2021). To successfully execute projects, project management employs a variety of tactics, methodologies, and project management best practices. As part of this, plans must be created, time and money must be managed, quality standards must be met, resources and communications must be managed, risks must be identified, procurement operations must be carried out, and stakeholders must be engaged (Kerzner, 2017; Kamaruzaman et al., 2019; Sherwani et al., 2020).

IR 4.0, shown in Figure 7.1, is a dynamically growing research field that integrates information from various academic disciplines to produce innovative manufacturing solutions (Qureshi et al., 2020). Technological advancements will have a significant impact on project management in the future. Artificial intelligence (AI) will progress from simple task automation to predictive project analytics, advice, and actions (Tsaramirsis et al., 2022).

Project teams are more successful than ever before, even though the capabilities of new technology frequently raise the expectations

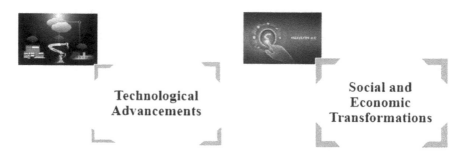

Technological Advancements

Social and Economic Transformations

FIGURE 7.1   Industrial Revolution (IR 4.0).

of customers and top management. As a result, project managers gain countless advantages, ranging from simpler communication to more precise reporting via emerging technologies (Maskuriy et al., 2019; Teo et al., 2021). The discussion of each knowledge area, along with its enhancement through IR 4.0, is discussed subsequently.

7.2 PROJECT INTEGRATION MANAGEMENT

All other project management knowledge areas fall under the purview of project integration management. Individual procedures and tasks are woven together into a single project with clear objectives and deliverables. Project integration management is achieved using various tools and techniques, including project charters, project plans, change management, project status reports, and project management software (Demirkesen and Ozorhon, 2017). Project integration management is critical to the success of any project.

Effective project integration management requires careful planning, attention to detail, and the use of appropriate tools and techniques. The benefits of successful project integration management include improved project outcomes, increased stakeholder satisfaction, and improved project team morale. It keeps the process moving along smoothly and effectively. Project management requires the management of project integration (Langston, 2013). It entails controlling the procedures used to combine diverse project components. This refers to both the project's physical and digital components as well as the individuals participating. It also requires managing relationships between stakeholders and handling changes that arise during the project (Hardin and McCool, 2015).

7.2.1 Conventional Way of Dealing

The traditional method of project integration management entails several connected procedures intended to harmonize every facet of a project. To establish goals and stakeholders, a project charter is first created. Then, a comprehensive project management plan is created, including subordinate plans for various areas and methods for execution, monitoring, and closure. The project work is subsequently completed while following the plan's instructions. Changing management procedures ensures that modifications are handled methodically. Validating the scope and maintaining control over expenses and timelines ensure that the original plans are followed. To uphold standards and deal with uncertainties, quality assurance and risk management are essential. For efficient collaboration, communication, procurement, and stakeholder engagement are constantly monitored. Finally, the project or project phase is formally concluded, enabling time for reflection and knowledge sharing. The project manager is essential to coordinating these steps, coordinating efforts, and assuring the project's success in the end.

7.2.2 Advancements through IR 4.0

Project integration management could undergo a revolution because Industry 4.0 introduced cutting-edge techniques and technology. Data analytics, automation, and the integration of digital tools can help projects succeed and operate more effectively. Accurate progress tracking and prediction insights are made possible by real-time data integration and digital project management platforms powered by artificial intelligence. Blockchain enables transparent and secure record-keeping, while collaborative virtual environments promote fluid international collaboration.

Risk anticipation is aided by predictive analytics, communication is streamlined by automated reporting, and people and materials are best utilized by intelligent resource allocation. Change is welcomed by agile approaches, and digital twin technology provides accurate simulations. Robotic process automation minimizes errors, and smart contracts automate operations. Continual feedback loops and monitoring guarantee adaptable strategies. Project integration management may use the potential of IR 4.0 to improve coordination, efficiency, and results across projects of all sizes and complexity by embracing these innovations, as shown in Figure 7.2.

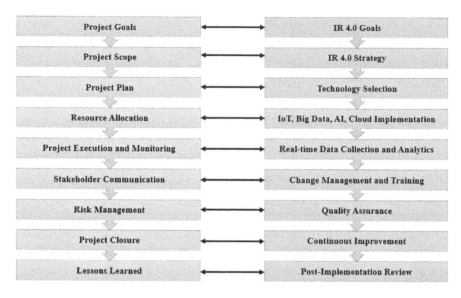

FIGURE 7.2   Integration of Project Integration Management and IR 4.0.

Example 7.1

A manufacturing business launches an Industry 4.0 project to streamline its production procedures and boost productivity. The project's goal is to connect the production environment by fusing IoT sensors, data analytics, and automation. Specific tasks are determined during the planning stage, such as integrating data analytics platforms, deploying IoT sensors, and putting automation solutions in place. It is decided how different jobs are dependent on one another, such as how seamlessly sensor data is integrated into the analytics platform. The development of a thorough project schedule takes resources and dependencies into account. Estimated costs consider expenditures for IoT sensors, analytics programs, and essential infrastructure changes. Risks are evaluated, and mitigation plans are presented, including data security and compatibility concerns. IoT sensors are deployed, data is integrated, and automation is put into place throughout execution. Employee training is provided, and development is tracked. When a project is finished, success criteria, such as increased productivity and production efficiency, are assessed, and lessons learned are recorded. The use of project integration management concepts in effectively implementing an Industry 4.0 effort is illustrated by this numerical example, which ensures smooth coordination of numerous stakeholders and technology.

EXAMPLE 7.2 THOUGHT BOX FOR STUDENTS

How does the integration of Industry 4.0 technologies impact the coordination and communication among project stakeholders?

7.3 PROJECT SCOPE MANAGEMENT

Project scope management ensures that project goals are met within the restrictions of time and resource constraints. This includes creating a project plan, figuring out what must be done and in what order, calculating costs, and monitoring progress concerning the plan (Moustafaev, 2014; Abdilahi et al., 2020). Effective project scope management helps to ensure that the project stays on track and within budget. It also helps to ensure that the project team has a clear understanding of the project goals and objectives, which helps to improve project performance and reduce project risks (Al-Rubaiei et al., 2018).

Effective scope management requires strong communication, collaboration, and stakeholder engagement. To effectively manage project scope, project managers should establish clear project goals and objectives, define project requirements and deliverables, and establish a process for managing scope changes. They should also ensure that stakeholders are engaged throughout the project and that any changes to the project scope are properly documented, approved, and managed.

7.3.1 Conventional Way of Dealing

In traditional project scope management, project boundaries are defined, managed, and controlled methodically. To understand the needs and aspirations of the stakeholders, requirements are first gathered from them. Then, a thorough and lucid scope statement is created by analyzing, documenting, and synthesizing these needs. After that, the scope is further defined and divided into work packages as part of the development of the work breakdown structure (WBS), which hierarchically arranges tasks. Scope adjustments are assessed against the baseline as the project moves forward to determine their need and impact. Careful change control is used throughout this procedure to avoid scope creep. Verifying the scope ensures that project deliverables adhere to the original specifications and that any deviations are corrected. The project manager is responsible for managing these procedures and working with stakeholders to ensure a precise and narrow project scope. This reduces the risk of scope-related problems and increases the possibility that the project will succeed.

7.3.2 Advancements through IR 4.0

Industry 4.0's growth brings new opportunities for improving project scope management through creative reforms. Project scope can be established, tracked, and managed more precisely by utilizing digital technology and data-driven methodologies. Real-time and historical project data can be thoroughly analyzed using advanced data analytics and artificial intelligence, allowing for more accurate scope identification and requirement collecting. Stakeholders may see project concepts and design thanks to virtual reality and augmented reality technologies, which promote a better understanding and minimize scope misunderstandings.

With the help of blockchain technology, change management is secure, and transparent, and tracks any modification to the project's scope. Additionally, distributed teams' communication and coordination are improved by collaborative digital platforms, which helps teams' scope definition and alignment. Project scope management can achieve increased effectiveness by incorporating these changes inspired by IR 4.0. Project scope management may improve accuracy, decrease scope creep, and increase stakeholder engagement throughout a project by incorporating several IR 4.0–inspired innovations, as shown in Figure 7.3.

7.4 PROJECT TIME MANAGEMENT

Effective project time management helps to ensure that the project is completed on time, within budget, and with the desired outcomes. It also helps to identify potential risks and delays early in the project, which allows project managers to take proactive steps to mitigate

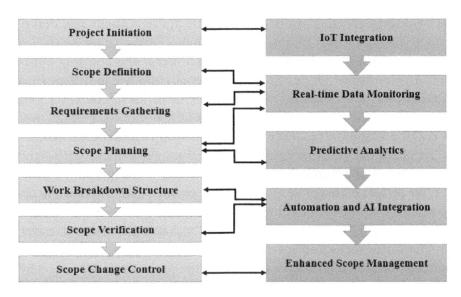

FIGURE 7.3   Integration of Project Scope Management and IR 4.0.

these risks and minimize delays (Chin and Hamid, 2015). To effectively manage project time, project managers should establish a detailed project plan that identifies the activities required to deliver the project objectives, the dependencies between these activities, and the estimated duration of each activity. They should also establish a clear schedule that defines the start and end dates for each activity and develop a process for monitoring and controlling the project schedule (Gładysz et al., 2015).

Effective project time management requires strong communication, collaboration, and stakeholder engagement. Project managers should ensure that stakeholders are engaged throughout the project and that any changes to the project schedule are properly documented, approved, and managed. Project time management is a critical process that helps to ensure the success of a project. Effective time management requires a detailed understanding of project requirements, strong communication, and stakeholder engagement. By effectively managing project time, project managers can ensure that the project stays on track, within budget, and delivers the desired outcomes.

To use cutting-edge technologies to streamline its production processes, a manufacturing company launches an Industry 4.0 project. Implementing IoT sensors, data analytics, and automation technologies are all part of the project. Specific tasks are determined during project planning, such as integrating data analytics platforms, installing IoT sensors on machinery, and creating automation solutions. The project manager divides these duties into smaller ones and gives them reasonable deadlines. The start and finish dates for each job are specified in a project schedule that is created based on resource availability and task dependencies.

To guarantee on-time project completion, the critical path—the longest series of dependent tasks—is identified. Throughout the project, the project manager keeps a close eye on each task's advancement, compares it to the original timeline, and quickly resolves any delays. Automated data gathering and analysis technologies make it easier to spot bottlenecks and inefficiencies, which empowers the project manager to decide how to best advance the project's timeframe. The organization successfully adopted Industry 4.0 technology within the anticipated timeframe thanks to good project time management, which boosts production efficiency and decreases downtime.

7.4.1 Conventional Way of Dealing

Traditional project time management includes a planned method for setting up, keeping track of, and managing a project's timeline. It starts with the development of a project schedule, which is accomplished through the steps of task definition, time estimation, and logical sequencing. This leads to a project schedule that specifies the timing of each task. As the project moves forward, the project manager keeps a careful eye on how things are going concerning the schedule, to spot any delays or deviations. Regular evaluations enable modifications, including rearranging work or allocating more resources to deal with bottlenecks. The minimum timetable for the project is determined by the longest sequence of tasks, which is identified using critical path analysis. These techniques allow for the management of time-related hazards and the implementation of necessary remedial measures. By monitoring time management, the project manager may make sure that everything moves along smoothly as it approaches completion, including keeping the project on schedule and meeting deadlines.

7.4.2 Advancements through IR 4.0

Project time management has the potential to be transformed by Industry 4.0, which ushers in developments that streamline

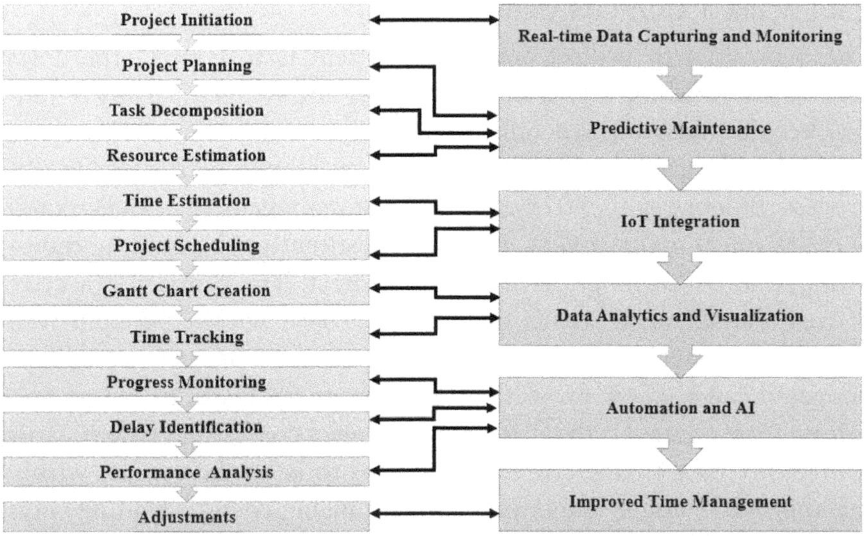

FIGURE 7.4   Integration of Project Time Management and IR 4.0.

scheduling, efficiency, and precision. Project timeframes can be continuously monitored via real-time data integration from IoT devices and smart sensors, delivering quick updates on progress and any delays. AI-powered predictive analytics can spot trends and anticipate future stumbling blocks, allowing for proactive adjustments to timetables. Remote teams may work together smoothly in collaborative virtual environments, which speeds up decision-making and lowers communication lag. Routine chores can be accelerated via robotic process automation, freeing up time for important project operations. Digital twins also give you the capacity to mimic project processes and foresee time-related problems before they arise. Project time management can achieve previously unseen levels of agility, reactivity, and accuracy by adopting these IR 4.0–inspired changes, which will ultimately lead to on-time project delivery and increased stakeholder satisfaction, as shown in Figure 7.4.

EXAMPLE 7.3 THOUGHT BOX FOR STUDENTS

How does the utilization of real-time data from connected devices in Industry 4.0 influence project scheduling and time management processes?

7.5 PROJECT COST MANAGEMENT

The main goal of project cost management is to control the costs involved in finishing a project on schedule and budget. Setting a budget, keeping tabs on expenditures associated with the project, and finding strategies to maximize resources are all necessary steps in this process (Blocher et al., 2019). Project cost management is an essential part of project management. It involves estimating, budgeting, monitoring, and controlling the costs of a project. The primary objective of cost management is to complete the project within the approved budget. Cost management is crucial because if the project exceeds the budget, it can result in project failure or financial losses to the organization. Cost control is the final step in project cost management. Cost control involves taking corrective actions to keep the project within the approved budget. This can include reducing costs or finding ways to complete the project more efficiently. Cost control is essential to ensure that the project is completed within the approved budget.

Several tools and techniques can be used for cost management. One such technique is earned value management (EVM). EVM is a project management tool that combines schedule and cost performance data to help project managers measure project progress and forecast project completion dates and costs. EVM is a useful technique for cost management because it provides a single measure of project performance, which can be used to identify potential problems and take corrective actions. Another useful tool for cost management is cost-benefit analysis (CBA).

CBA is a technique for evaluating the costs and benefits of a project (Smith, 2014). CBA helps project managers determine whether a project is worth pursuing based on its potential benefits compared to its costs. Hence, project cost management is an essential part of project management. It involves estimating, budgeting, monitoring, and controlling the costs of a project. Accurate cost estimation and budgeting are essential to ensure that the project is completed within the approved budget. Cost monitoring and control are essential to identify and correct any deviations from the budget. Various tools and techniques, such as earned value management and cost-benefit analysis, can be used for effective cost management (Smith, 2016; Hansen et al., 2021).

7.5.1 Conventional Way of Dealing

In traditional project cost management, the financial resources needed for a project are planned, estimated, budgeted, monitored, and controlled. By evaluating the costs associated with each activity and resource required for the project, a project budget is first created. These projections are adjusted as the project develops, to account for variables like inflation, scope changes, and unforeseen costs. The project manager continuously compares actual expenses to the budget as the project progresses, to spot any discrepancies.

The impact and causes of variations are then determined through analysis. To manage budgetary deviations, cost control measures are put in place. These steps may comprise resource reallocation, plan revisions, or the search for cost-saving options. The performance of the project is evaluated in terms of cost-effectiveness and advancement using techniques like earned value management. The control of the project manager over financial resources ensures efficient management, reducing budget overruns, optimizing resource allocation, and boosting the project's financial performance.

7.5.2 Advancements through IR 4.0

Project cost management could undergo a radical transformation with the introduction of innovations that improve accuracy, transparency, and efficiency thanks to Industry 4.0. AI-driven algorithms can offer detailed cost projections and insights based on historical data and ongoing project performance by integrating real-time data analytics. Advanced cost estimation software can use information from related projects to produce more accurate forecasts.

Blockchain technology ensures safe and impenetrable cost tracking, boosting stakeholders' responsibility and trust. Robotic process automation also simplifies financial reporting and typical cost calculations, lowering administrative burden and manual error. Digital platforms that encourage collaboration make it easier for teams to share information and communicate efficiently, ultimately improving cost management. Project cost management can achieve increased precision, cost optimization, and budget adherence by adopting these changes motivated by IR 4.0, which will ultimately result in successful project results and customer satisfaction, as shown in Figure 7.5.

7.6 PROJECT QUALITY MANAGEMENT

Project quality management is a crucial aspect of project management that aims to ensure that the project meets the quality standards and requirements of the stakeholders. It involves planning, executing, and controlling quality-related activities throughout the project lifecycle (Baker, 2018). Several tools and techniques can be used for project quality management. One such technique is Six Sigma, which is a data-driven methodology for improving quality and reducing defects in processes. Six Sigma involves using statistical analysis to identify and eliminate sources of variation in processes, resulting in improved quality outcomes and reduced costs.

Another useful tool for quality management is quality function deployment (QFD). QFD is a technique for translating customer needs and requirements into specific technical requirements for the project. QFD involves using a matrix to identify the relationships between customer needs, technical requirements, and project deliverables, ensuring that the project meets the customer's expectations. Project quality management is a critical aspect of project management

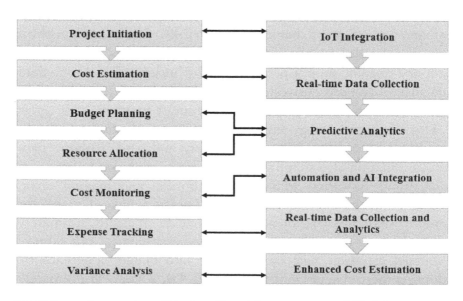

FIGURE 7.5   Integration of Project Cost Management and IR 4.0.

that aims to ensure that the project meets the quality standards and requirements of the stakeholders. By implementing a robust quality management process, project teams can deliver high-quality projects that meet the expectations of the stakeholders and contribute to the success of the organization (Alwaly and Alawi, 2020).

7.6.1 Conventional Way of Dealing

Traditional project quality management is a systematic method for organizing, carrying out, and keeping track of tasks that guarantee the project's deliverables adhere to predetermined quality standards. It starts with creating a quality management plan that specifies the project's quality goals, standards, and strategies for achieving them. The project team adheres to specified quality processes while carrying out the project, using techniques like inspections, audits, and reviews to spot and address any deviations from the established quality standards.

Key performance indicators are continuously monitored and measured to determine how well the project is adhering to its quality objectives. If problems do occur, steps are taken to fix them and keep them from happening again. The improvement of current quality practices also makes use of the lessons learned from earlier initiatives. In the end, the project manager is essential in ensuring that the project's results meet the established quality requirements, encouraging stakeholder satisfaction, and minimizing the risks associated with subpar quality.

7.6.2 Advancements through IR 4.0

With the introduction of innovations that raise standards, consistency, and overall project excellence, the rise of Industry 4.0 has the potential to fundamentally alter project quality management. Real-time data gathering is made simple by connecting IoT devices and sensors, enabling continuous monitoring of important quality indicators. These data can be analyzed by sophisticated machine learning and data analytics algorithms to forecast possible quality problems and suggest mitigation measures.

Real-time inspections and simulations are made possible by collaborative virtual environments, allowing stakeholders to spot and resolve quality issues early on. Transparency and accountability are improved across the supply chain thanks to blockchain technology, which guarantees the immutability of quality records. Digital twin technology also enables extensive testing and optimization in a risk-free virtual environment, supporting superior quality results. Project quality management may raise its standards, reduce errors, and foster a culture of continuous improvement by adopting these changes motivated by IR 4.0, leading to goods and services that meet or exceed client expectations, as shown in Figure 7.6.

Example 7.4

A manufacturing company launches an Industry 4.0 project to use cutting-edge technologies to improve the quality of its production operations. IoT sensors, data analytics, and automated inspection systems will all be used in this project. Specific quality goals, such as lowering faults and ensuring uniform product quality, are set during project planning. Quality measurements are established, like defect rates and customer satisfaction scores, to help achieve these goals. The project team creates a thorough quality management plan that details quality control measures, such as inspection techniques, sampling strategies, and data analysis methodologies. IoT sensors are installed to continuously collect data for analysis while monitoring important production indicators in real time. Tools for data analytics are used to spot abnormalities, pinpoint quality problems, and launch corrective measures. To assure consistent product quality and reduce human mistakes, automated

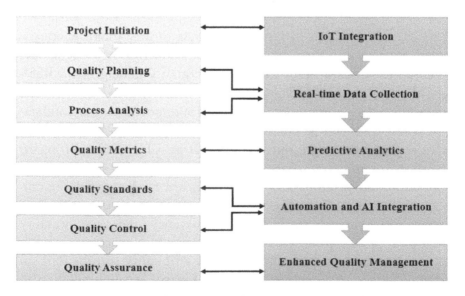

FIGURE 7.6 Integration of Project Quality Management and IR 4.0.

inspection methods are used. The project team performs routine quality audits throughout project execution to evaluate adherence to quality standards and implement continuous improvement strategies. In the Industry 4.0 environment, the organization successfully fulfills its quality objectives, leading to improved product quality, fewer defects, and increased customer happiness.

7.7 PROJECT HUMAN RESOURCE MANAGEMENT

The primary objective of project human resource management is to ensure that the project team has the necessary skills, knowledge, and experience to deliver the project successfully. The project manager should also develop a plan for managing the project team's performance. This involves establishing performance metrics and performance standards that can be used to evaluate the project team's performance (Zaouga et al., 2019).

The project manager should regularly review the team's performance against these metrics and standards and take corrective action as necessary. Effective communication is essential for project human resource management. The project manager should ensure that there is open and effective communication between the project team members and stakeholders. This can involve regular team meetings, status reports, and progress updates. The project manager should also ensure that there is effective communication between the project team and external stakeholders, such as clients and suppliers. Finally, project human resource management also involves managing the project team's conflicts and challenges. Conflict can arise due to differences in opinions, values, or working styles (Zaouga et al., 2019).

The project manager should ensure that conflicts are identified and addressed promptly to prevent them from escalating and affecting the project's success. The project manager should also be aware of the challenges faced by the project team and take steps to address them, such as providing additional resources or training. Project human resource management is a critical aspect of project management that focuses on the people involved in the project. It includes the processes and activities needed to identify, acquire, develop, and manage the project team. Effective project human resource management involves identifying the roles and responsibilities needed for the project, acquiring the project team, developing, and managing the team, managing the team's performance, and managing conflicts and challenges. By implementing an effective project human resource management plan, project managers can ensure that the project team has the necessary skills, knowledge, and experience to deliver the project successfully (Carden et al., 2021).

7.7.1 Conventional Way of Dealing

A systematic method is used in traditional project human resource management to efficiently manage the personnel participating in a project. It starts by outlining the roles and responsibilities needed for the project as well as the knowledge and abilities required for each function. Following that, team members are hired or assigned based on these criteria. Through exercises like team building, training, and developing effective communication channels, the project team is established.

The project manager keeps an eye on the team's performance as it develops and handles any disputes, worries, or problems that may come up. Team members are held accountable for exceeding expectations and contributing to the success of the project through regular feedback and performance reviews. The project manager is crucial in boosting the team's morale and encouraging cooperation. Effective project human resource management ultimately makes sure that the right people are in the correct roles, cooperating to accomplish project goals and enhance overall performance.

7.7.2 Advancements through IR 4.0

Project HRM will experience a dramatic change because of the introduction of changes that will improve teamwork, skill development, and workforce productivity. Teams may easily interact across regional borders thanks to the integration of cutting-edge communication tools and digital platforms, which promotes global cooperation and information sharing. Systems for managing talent that are AI-powered can find skills shortages and provide individualized learning routes to encourage upskilling and continual learning. Technologies like virtual reality and augmented reality can enhance immersive training experiences, enabling workers to learn useful skills in a risk-free setting.

Additionally, proactive workforce planning is made possible by predictive analytics, which can foresee resource requirements and indicate prospective staffing concerns. Blockchain technology can increase openness in hiring and performance management procedures, resulting in objective and fair assessments. Project human resource management can foster a dynamic, adaptable workforce, encourage skill development, and maximize the potential of human capital by adopting these IR 4.0–inspired changes, which will ultimately result in more successful project outcomes and higher employee satisfaction, as shown in Figure 7.7.

7.8 PROJECT COMMUNICATION MANAGEMENT

All stakeholders are informed of the project's status and have timely access to the information they require, thanks to effective project communications management. This includes creating a communication plan, establishing channels for communication, successfully interacting with team members and other stakeholders, and handling stakeholder conflicts (Samáková et al., 2013). Project communication management is a crucial aspect of project management that involves the effective planning, execution, and control of communication within a project. It ensures that the right information is delivered to the right people at the right time to support project success. Effective communication management enables stakeholders to understand the project's objectives, progress, risks, and issues and enables them to make informed decisions (Samáková et al., 2013).

Effective project communication management also requires the use of appropriate communication channels. The communication channels used should be determined by the project's needs and stakeholder preferences. Examples of communication channels include email, video

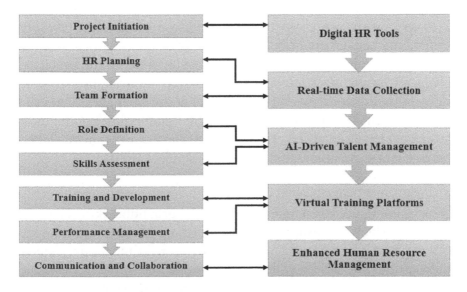

FIGURE 7.7   Integration of Project Human Resource Management and IR 4.0.

conferencing, instant messaging, social media, and face-to-face meetings. The project manager should ensure that the selected communication channels are reliable, secure, and accessible to all stakeholders. Regular stakeholder meetings and reporting are also essential for effective project communication management. The project manager should provide regular progress reports to stakeholders, including updates on project milestones, risks, issues, and changes. The project manager should also ensure that stakeholders can provide feedback on the project's progress and make suggestions for improvement. Effective project communication management also requires strong leadership and interpersonal skills. The project manager should be able to communicate clearly and effectively, listen actively, and be able to resolve conflicts that may arise during the project (Mikhieieva and Waidmann, 2017).

7.8.1 Conventional Way of Dealing

Typical project communication management takes a methodical approach to organizing, carrying out, and keeping track of project-related communications. It starts by analyzing the information needs of different stakeholders, and their communication needs, and setting up specific communication goals. The next step is to create a communication management plan that specifies who will receive what information, how it will be provided, and how frequently.

The project manager helps team members, partners, and stakeholders communicate clearly as the project moves forward. To keep everyone up to date on project developments, regular status meetings, progress reports, and updates are held. Rapid communication of any problems, modifications, or hazards enables quick decision-making and problem-solving. To ensure that communication channels are open, transparent, and consistent throughout the project, the project manager is crucial. Through open, transparent, and consistent communication channels throughout the project lifecycle, the project manager promotes collaboration, aligns expectations, and reduces misunderstandings.

7.8.2 Advancements through IR 4.0

Project communication management is given groundbreaking chances by IR 4.0, which also ushers in changes that improve communication, collaboration, and openness. Teams may use real-time texting, video conferencing, and virtual collaboration platforms to interact without difficulty regardless of location by integrating modern communication tools. With the use of AI-driven sentiment analysis, project managers

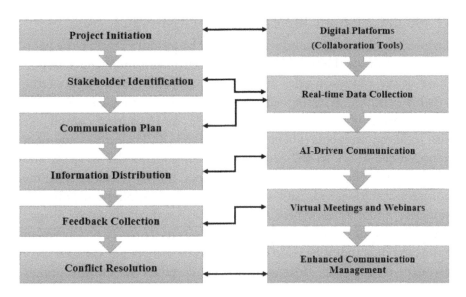

FIGURE 7.8 Integration of Project Communication Management and IR 4.0.

can determine how stakeholders will respond to project updates and adjust their communication plans accordingly. Immersive meetings and design reviews are made possible by collaborative virtual environments, encouraging more engaging and participatory discussions.

Additionally, crucial project communications can be protected and authenticated using blockchain technology, confirming their accuracy and source. Natural language processing and automated chatbots streamline information retrieval and inquiry resolution, improving timeliness. Project communication management may maximize stakeholder participation, reduce misconceptions, and promote better relationships with stakeholders by adopting these changes motivated by IR 4. 0. Project communication management may improve stakeholder participation, reduce misconceptions, and promote an open and effective communication culture by adopting these IR 4.0–inspired reforms, which will ultimately result in better decisions and successful project outcomes, as shown in Figure 7.8.

EXAMPLE 7.5 THOUGHT BOX FOR STUDENTS

How can one elaborate the impact of remote work enabled by Industry 4.0 on team collaboration, communication, and skill development within project teams?

7.9 PROJECT PROCUREMENT MANAGEMENT

A project's commodities and services must be procured for it to be completed, and this is the focus of project procurement management. This phase of the PM knowledge areas covers everything from discovering possible vendors, assessing supplier offers, negotiating contracts and terms, managing the procurement process, and ensuring compliance with legal requirements (Pheng and Pheng, 2018). Project procurement management refers to the process of acquiring goods, services, and other resources needed to complete a project. It involves identifying the requirements for goods and services, selecting suppliers, negotiating contracts, and managing procurement activities throughout the project lifecycle. One of the primary benefits of effective project procurement management is that it helps to ensure that the project has the necessary resources to complete it successfully. By carefully selecting suppliers and negotiating favorable contracts, project managers can ensure that they have access to the resources they need to complete the project on time and within budget (Gutierrez Aguilar, 2022).

Effective project procurement management also helps to minimize the risk of cost overruns and delays. By carefully planning and managing procurement activities, project managers can identify potential problems early in the project and take corrective action before they become more serious. Another benefit of effective project procurement management is that it helps to ensure that the project complies with relevant regulations and laws. Procurement regulations vary depending on the industry and location, and failure to comply with them can result in legal and financial penalties. By understanding and complying with procurement regulations, project managers can minimize the risk of legal and financial problems (Alwaly and Alawi, 2020).

7.9.1 Conventional Way of Dealing

Traditional project procurement management entails a methodical approach to obtaining the materials from outside sources that are required to support project execution. It starts with determining what needs to be acquired and whether to make or purchase the necessary items. These choices inform the development of procurement strategies, which may involve actions like negotiated or competitive contracts.

The project team then draughts procurement documents outlining the criteria and criteria for prospective suppliers or vendors.

To choose the best vendor, bids or proposals are requested, and a selection procedure is used to make the decision. A formal contract is established after a vendor has been selected, outlining the terms, conditions, and deliveries. The project manager is responsible for supervising the procurement process and making sure that contracts are carried out as intended and that vendors fulfill their obligations. Contract performance is tracked throughout the project, and any alterations or conflicts are handled following the specified protocols. Effective project procurement management makes it possible to acquire the resources that are required promptly and affordably while reducing the risks connected to external dependencies.

7.9.2 Advancements through IR 4.0

Project procurement management will undergo fundamental modifications because of the rise of Industry 4.0, which will improve effectiveness, transparency, and strategic sourcing. Artificial intelligence (AI)–driven procurement systems can analyze historical data and market trends through the integration of digital platforms to streamline supplier selection and contract negotiation procedures. By guaranteeing safe and impenetrable records of procurement transactions, blockchain technology promotes accountability and confidence.

By predicting procurement demands, advanced data analytics can enable proactive inventory management and shorten lead times. Real-time contact between suppliers and customers is made possible by collaborative digital platforms, which also streamline information sharing and order tracking. Smart contracts can also automate the workflows involved in procurement, ensuring that the terms and conditions are followed. Project procurement management can achieve cost savings, reduce risks, and strengthen supplier relationships by adopting these IR 4.0–inspired reforms, ultimately resulting in effective project results and optimal resource utilization, as shown in Figure 7.9.

7.10 PROJECT RISK MANAGEMENT

The main goal of project risk management is to identify potential risks or issues that could delay the project and to develop plans or strategies to address them if they materialize. The Project Management Body of Knowledge (PMBOK) states that this comprises doing a quantitative risk analysis of the project's risks, creating backup plans in case those

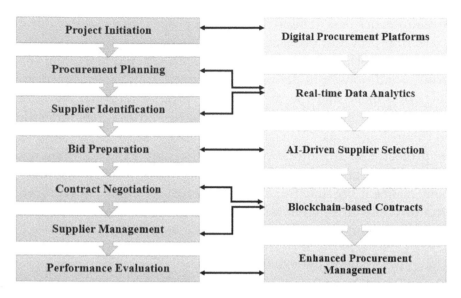

FIGURE 7.9   Integration of Project Procurement Management and IR 4.0.

risks come to pass, and keeping an eye on risks throughout the project's lifecycle (Willumsen et al., 2019). Project risk management is the process of identifying, assessing, and managing risks that may impact a project's success. It involves identifying potential risks, assessing their likelihood and impact, developing strategies to mitigate or avoid them, and monitoring and controlling risks throughout the project lifecycle (Taghipour et al., 2015).

One of the primary benefits of effective project risk management is that it helps to minimize the risk of project failure. By identifying and addressing potential risks early on in the project, project managers can reduce the likelihood of cost overruns, delays, and other problems that can jeopardize project success. Effective project risk management also helps to improve project planning and decision-making. By considering potential risks and their potential impact on the project, project managers can make more informed decisions about project scope, objectives, and resources. This can help to ensure that the project is more likely to meet stakeholder expectations and deliver the desired results. Another benefit of effective project risk management is that it helps to build stakeholder confidence in the project's success. By demonstrating that potential risks have been identified and addressed, project managers can help build trust and confidence in the project's success among stakeholders (El Yamami et al., 2017).

7.10.1 Conventional Way of Dealing

In traditional project risk management, potential uncertainties that might affect project goals are identified, evaluated, and managed systematically. Starting with brainstorming sessions, historical data analysis, and advice from experts, potential risks are identified.

The possibility of these risks and their potential effects on the project's scope, schedule, budget, and quality are then evaluated. Following analysis, methods for mitigating or addressing identified risks are established. These tactics could include risk acceptance, transfer, minimization, or avoidance. Throughout the project, the project manager closely monitors and tracks these risks, modifying response plans as necessary as new facts or circumstances arise. The project manager strives to reduce risks' negative effects on the project's success by proactively controlling them. To guide future initiatives, lessons learned from addressing hazards are also documented. A project's ability to negotiate uncertainties, make the best decisions, and increase the likelihood of reaching desired objectives are all improved by competent project risk management.

7.10.2 Advancements through IR 4.0

Industry 4.0 opens a world of revolutionary possibilities for PRM, bringing about changes that support proactive risk detection, assessment, and mitigation. Risk variables can be continuously monitored, enabling the early identification of possible risks, with the help of real-time data integration from IoT devices and sensors. Utilizing previous project data and outside variables, predictive analytics powered by AI algorithms can foresee new risks and facilitate prompt risk response planning. Scenario planning and risk simulations are made easier by collaborative virtual environments, enabling teams to assess the effects of different risk scenarios. By ensuring transparent and unchangeable records of risk assessments and mitigation plans, blockchain technology improves accountability and compliance. Furthermore, regular risk management processes can be streamlined by robotic process automation, freeing up resources for strategic risk analysis. Project risk management can adopt these IR 4.0–inspired changes to become more proactive, adaptive, and efficient in navigating uncertainties, which will ultimately result in more robust projects and better outcomes, as shown in Figure 7.10.

7.11 PROJECT STAKEHOLDER MANAGEMENT

Effective project stakeholder management requires a range of skills and tools, including stakeholder analysis techniques, communication strategies, and conflict resolution skills. Project managers must also be able to balance the project's objectives with the needs and expectations of stakeholders to ensure that the project is successful and meets stakeholder expectations. One of the primary benefits of effective project stakeholder management is that it helps to ensure that stakeholders are engaged and committed to the project's success. By involving stakeholders in decision-making processes and addressing their concerns and objections, project managers can build trust and confidence in the project among stakeholders.

Effective project stakeholder management also helps to minimize the risk of project failure. By identifying potential stakeholder conflicts and concerns early in the project, project managers can take proactive measures to address these issues and minimize their impact on the project's success. Another benefit of effective project stakeholder management is that it helps to improve project

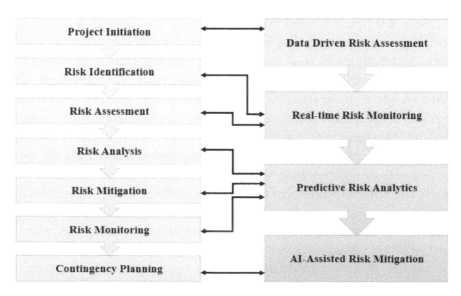

FIGURE 7.10 Integration of Project Risk Management and IR 4.0.

outcomes. By considering the needs and expectations of stakeholders throughout the project lifecycle, project managers can ensure that the project is aligned with stakeholder expectations and is more likely to deliver the desired results (Pandi-Perumal et al., 2015).

Hence, project stakeholder management is a critical aspect of project management. It involves identifying, analyzing, and managing stakeholders and their interests in a project. Effective project stakeholder management helps to ensure that stakeholders are engaged and committed to the project's success, minimize the risk of project failure, and improve project outcomes (Nauman and Piracha, 2016). Project managers must have a range of skills and tools to manage stakeholders effectively, including stakeholder analysis techniques, communication strategies, and conflict resolution skills. By effectively managing project stakeholders, project managers can increase the likelihood of project success and deliver projects that meet stakeholder expectations (Chung and Crawford, 2016).

7.11.1 Conventional Way of Dealing

In traditional project stakeholder management, the needs and expectations of all people or groups impacted by the project are identified, engaged with, and addressed using a structured method. It begins with identifying important stakeholders and gaining an understanding of their motivations, issues, and impact on the project's results. The project team makes sure that stakeholders are aware of the project's goals, status, and potential effects through effective communication.

To involve stakeholders in decision-making and get their feedback throughout the project, engagement techniques are devised. Stakeholder relationships are actively managed by the project manager, who addresses disagreements, resolves problems, and modifies strategy in response to stakeholder input. The project manager makes sure that stakeholder expectations and project outcomes remain in sync as it progresses and that any adjustments or deviations are reported right away. Project stakeholder management is crucial in promoting collaboration, reducing opposition, and maximizing support for the project's objectives because the success of the project is intimately related to the happiness of stakeholders.

7.11.2 Advancements through IR 4.0

Industry 4.0's development offers a revolutionary perspective for project stakeholder management, bringing about changes that improve collaboration, participation, and transparency. Stakeholders can easily communicate through real-time texting, video conferencing, and virtual collaborative spaces using advanced digital communication systems, overcoming geographic obstacles. Project managers can successfully modify their communication tactics by using AI-driven sentiment analysis to assess the sentiments of their stakeholders.

Stakeholders can engage in immersive interactions with project data, designs, and progress thanks to collaborative virtual environments, which promote greater understanding and participation. Blockchain technology can authenticate stakeholder interactions and secure them, promoting confidence and transparency in the exchange of information. Personalized engagement initiatives can be guided by data analytics, which can also reveal insights into stakeholder preferences and expectations. Project stakeholder management may improve relationships, better manage expectations, and better match project goals with stakeholder demands by adopting these IR 4.0–inspired changes. This will ultimately increase project success and stakeholder satisfaction, as shown in Figure 7.11.

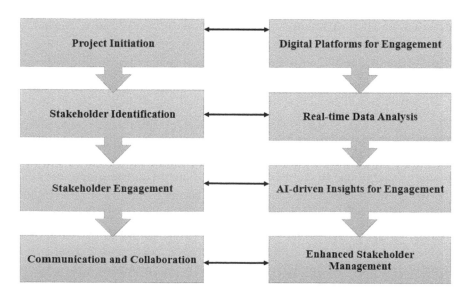

FIGURE 7.11   Integration of Project Stakeholder Management and IR 4.0.

EXAMPLE 7.6 THOUGHT BOX FOR STUDENTS

How does the utilization of social media analytics, sentiment analysis, and real-time feedback mechanisms introduced by Industry 4.0 impact the identification, engagement, and management of stakeholders in project management?

7.12 PROJECT SAFETY MANAGEMENT

Effective project safety management requires a range of skills and tools, including hazard and risk assessment techniques, safety planning and implementation strategies, and incident investigation and reporting procedures. Project managers must also be able to communicate effectively with project team members and other stakeholders to ensure that everyone understands the importance of project safety and their role in maintaining a safe working environment. One of the primary benefits of effective project safety management is that it helps to ensure the health and well-being of project team members and other stakeholders. By identifying and mitigating potential hazards and risks, project managers can reduce the likelihood of accidents or injuries on the project (Zachko et al., 2019).

Effective project safety management also helps to minimize project delays and costs associated with accidents or injuries. By implementing safety measures and conducting regular safety inspections, project managers can identify and address potential safety issues before they become major problems, minimizing the impact on project timelines and budgets. Another benefit of effective project safety management is that it helps to build trust and confidence among project team members and other stakeholders (Park and Huh, 2018). By demonstrating a commitment to project safety and taking proactive measures to ensure a safe working environment, project managers can build a culture of safety that promotes trust, collaboration, and teamwork.

Hence, project safety management is a critical aspect of project management. It involves identifying and mitigating potential hazards and risks associated with a project to ensure that the project is completed safely and without any accidents or injuries (Taghipour et al., 2015). Effective project safety management requires a range of skills and tools, including hazard and risk assessment techniques, safety planning and implementation strategies, and incident investigation and reporting procedures. By effectively managing project safety, project managers can ensure the health and well-being of project team members and other stakeholders, minimize project delays and costs, and build trust and confidence among project stakeholders.

7.12.1 Conventional Way of Dealing

An organized strategy for recognizing, mitigating, and assuring the safety of people and property connected to a project is part of conventional project safety management. It starts with a thorough analysis of potential threats to health and safety that might materialize throughout project activities. Then steps are taken to reduce or eliminate these risks, concentrating on both the physical security of the workforce and the defense of project-related assets. The project team, contractors, and other pertinent parties are informed of the specified safety rules, guidelines, and processes. To monitor adherence to safety standards and spot any deviations or dangers, routine safety inspections and audits are carried out. The project manager oversees implementing safety regulations, establishing a safety-conscious environment, and swiftly addressing any safety issues that come up throughout the project. The prevention of potential accidents or occurrences that can negatively impact the advancement, reputation, and general success of a project depends on ongoing examination and development of safety procedures.

7.12.2 Advancements through IR 4.0

The revolutionary innovations brought about by Industry 4.0 have the potential to completely transform project safety management, improving safety procedures, risk mitigation, and overall project success. Real-time data gathering enables continuous monitoring of safety parameters and quick response to possible threats through the integration of IoT devices and sensors. AI-powered predictive analytics may analyze past safety data and outside factors to foresee new safety concerns, enabling proactive safety measures. Collaborative virtual environments provide immersive safety training and simulations, providing employees with a secure setting in which to practice following procedures and responding to emergencies.

Wearable technology can offer in-the-moment notifications and analytics to make sure employees follow safety procedures. Additionally, blockchain technology can authenticate and safeguard safety compliance records, assuring accountability and transparency throughout the project. Project safety management may raise safety standards, reduce accidents, and foster a safety-conscious culture by adopting these changes inspired by IR 4.0, which will ultimately result in a safer workplace and better project outcomes, as shown in Figure 7.12.

Example 7.7

A manufacturing company launches an Industry 4.0 project to use cutting-edge technologies to improve safety measures inside its production processes. Implementing IoT sensors, real-time monitoring systems, and automatic safety processes are all part of the project. Specific safety goals, such as reducing workplace accidents and creating a safe working environment, are set during project planning. The project team undertakes a thorough safety assessment to meet these goals, identifying potential risks and hazards related to the adoption of new technology. They create a safety management strategy that contains emergency response processes, employee training requirements, and safety protocols. IoT sensors are set up to track the functioning of machinery and the state of

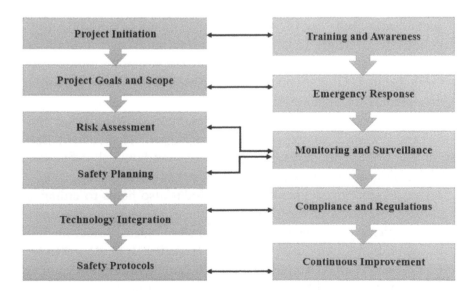

FIGURE 7.12   Integration of Project Safety Management and IR 4.0.

the environment, delivering real-time data for safety monitoring. Automated safety systems are used to spot dangerous situations, and sound alerts, and carry out prompt corrective actions.

Regular safety inspections are carried out during project execution to evaluate adherence to safety standards and pinpoint areas for improvement. To inform staff on how to use new technology and handle emergencies, the project team holds safety training sessions. The business successfully lowers workplace accidents, makes the workplace safer, and safeguards the well-being of employees in the Industry 4.0 environment by managing project safety properly.

7.13 PROJECT ENVIRONMENTAL MANAGEMENT

Project environmental management refers to the process of identifying and mitigating the environmental impacts of a project. The primary goal of project environmental management is to minimize or eliminate the negative environmental impacts of a project while maximizing its positive impacts.

Effective project environmental management requires a range of skills and tools, including environmental assessment techniques, mitigation planning and implementation strategies, and monitoring and reporting procedures. Project managers must also be able to communicate effectively with project team members and other stakeholders to ensure that everyone understands the importance of environmental management and their role in maintaining a sustainable project. One of the primary benefits of effective project environmental management is that it helps to minimize the negative impacts of the project on the environment. By identifying and mitigating potential environmental impacts, project managers can minimize or eliminate damage to ecosystems, reduce pollution, and conserve natural resources (Cardona-Meza and Olivar-Tost, 2017). Effective project environmental management also helps to enhance the positive impacts of the project on the environment. This may include implementing renewable energy technologies, protecting and restoring habitats, and using sustainable materials and practices. Another benefit of effective project environmental management is that it helps to build trust and confidence among project team members and other stakeholders. By demonstrating a commitment to environmental management and taking proactive measures to ensure a sustainable project, project managers can build a culture of environmental responsibility that promotes trust, collaboration, and teamwork (Hermano and Martín-Cruz, 2019).

Hence, project environmental management is a critical aspect of project management. It involves identifying and mitigating potential environmental impacts associated with a project to minimize or eliminate negative environmental effects while maximizing positive impacts. Effective project environmental management requires a range of skills and tools, including environmental assessment techniques, mitigation planning and implementation strategies, and monitoring and reporting procedures. By effectively managing the environmental impacts of a project, project managers can minimize damage to the environment, enhance positive impacts, and build trust and confidence among project stakeholders.

7.13.1 Conventional Way of Dealing

Conventional identification, evaluation, and mitigation of any potential environmental effects of project activities are all part of project environmental management. To identify potential environmental risks and issues related to the project, an environmental assessment is the first step. This evaluation assists in the creation of strategies and action plans to reduce adverse effects and advance sustainable practices. These tactics might include resource conservation, waste management, energy efficiency, and observance of pertinent environmental laws and standards. These methods are put into practice by the project team, which closely monitors environmental performance throughout the project's lifespan under the direction of the project manager. The effectiveness of mitigation measures is ensured through routine evaluations and audits, which also help assure compliance with environmental regulations. Project environmental management tries to do this by including environmental factors in project design and implementation. Project environmental management seeks to reduce ecological harm, encourage wise resource use, and help the project match sustainability objectives by incorporating environmental factors into project design and implementation.

7.13.2 Advancements through IR 4.0

Industry 4.0 introduces transformative reforms that hold the potential to reshape project environmental management, fostering sustainability, compliance, and eco-conscious practices. Through the integration of IoT sensors and data analytics, real-time environmental monitoring becomes feasible, enabling proactive identification of potential ecological impacts. AI-driven algorithms can analyze complex environmental data, aiding in predicting and addressing potential issues. Collaborative virtual environments offer a platform for stakeholders to assess and optimize environmental strategies, promoting a collective effort towards sustainable outcomes. Blockchain technology can ensure the transparency and traceability of environmentally sensitive processes, enhancing accountability and adherence to regulations. Moreover, digital twin technology allows for virtual simulations of environmental impact, enabling project teams to experiment with eco-friendly alternatives before implementation. By embracing these reforms inspired by IR 4.0, project environmental management can advance eco-conscious initiatives, minimize ecological footprints, and contribute to a greener, more sustainable future while achieving successful project outcomes, as shown in Figure 7.13.

7.14 PROJECT FINANCIAL MANAGEMENT

Project financial management is the process of planning, organizing, monitoring, and controlling financial resources to ensure the successful

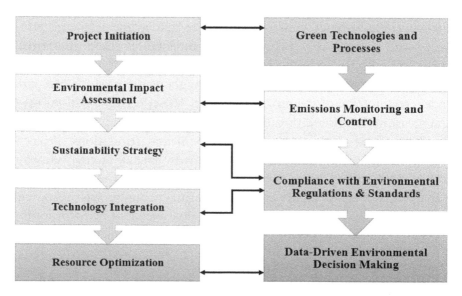

FIGURE 7.13   Integration of Project Environmental Management and IR 4.0.

completion of a project. The primary objective of project financial management is to ensure that the project is completed within the approved budget while achieving the desired outcomes. Effective project financial management requires a range of skills and tools, including financial planning and analysis, cost estimation, budgeting, resource management, and financial reporting (Jamali and Oveisi, 2016). Project managers must also be able to communicate effectively with project team members and other stakeholders to ensure that everyone understands the project's financial goals and their role in achieving them.

One of the primary benefits of effective project financial management is that it helps to ensure that the project is completed within the approved budget. This helps to minimize the risk of cost overruns and ensures that resources are used efficiently. Effective project financial management also helps to improve the project's overall financial performance. By monitoring the project's financial performance regularly, project managers can identify potential cost overruns or variances early and take corrective action to mitigate the risk. Another benefit of effective project financial management is that it helps to build trust and confidence among project stakeholders. By demonstrating a commitment to financial responsibility and taking proactive measures to ensure that the project stays within the approved budget, project managers can build a culture of financial discipline that promotes trust, collaboration, and teamwork. Project financial management is a critical aspect of project management (Alwaly and Alawi, 2020). It involves planning, organizing, monitoring, and controlling financial resources to ensure the successful completion of a project within the approved budget. Effective project financial management requires a range of skills and tools, including financial planning and analysis, cost estimation, budgeting, resource management, and financial reporting. By effectively managing the project's financial resources, project managers can minimize the risk of cost overruns, improve the project's overall financial performance, and build trust and confidence among project stakeholders (Rustamova, 2021).

7.14.1 Conventional Way of Dealing

Traditional project financial management uses a defined process for organizing, overseeing, and managing a project's financial components.

It starts with the creation of a project budget that specifies the anticipated expenditures for materials, activities, and deliverables. To ensure financial conformity, this budget acts as a reference point for the entire project. As the project moves forward, the project manager keeps a careful eye on budgeted expenses to spot any anomalies or variations. To manage budget overruns and maintain financial effectiveness, cost control measures are put in place. Stakeholders receive frequent financial reports and updates to keep them updated on the project's financial situation. The project manager works with the finance team to properly manage cash flow, allocate resources, and make decisions that are in line with the project's objectives. Project financial management ensures that financial resources are used effectively and that the project remains financially viable by upholding financial discipline and transparency, which contributes to the project's overall success.

7.14.2 Advancements through IR 4.0

Project financial management is presented with a paradigm shift by Industry 4.0, which introduces changes that improve financial accuracy, transparency, and strategic decision-making. Real-time financial monitoring is made feasible by the combination of sophisticated financial software and data analytics, enabling project managers to precisely track costs, income, and budget allocations. To forecast possible financial hazards and possibilities, AI-driven algorithms may analyze previous financial data and market patterns. This enables proactive financial planning. Integrated communication between financial stakeholders is made possible by collaborative digital platforms, which also promote quicker approvals and real-time financial reports. Financial records are safe and unchangeable thanks to blockchain technology, which also improves audit trails and compliance. Automated financial reporting also speeds up the creation of financial statements and predictions, minimizing human error. Project financial management can improve by adopting these changes motivated by IR 4.0. Project financial management may optimize resource allocation, lower financial risks, and support informed decision-making by adopting these IR 4.0–inspired changes, which will ultimately result in better financial outcomes and project success, as shown in Figure 7.14.

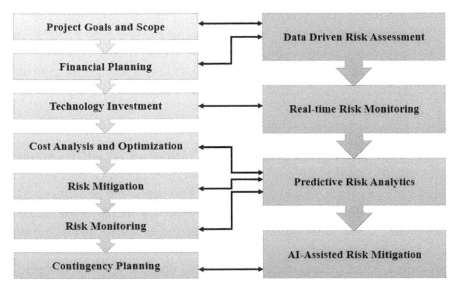

FIGURE 7.14   Integration of Project Financial Management and IR 4.0.

Example 7.8

A manufacturing corporation launches an Industry 4.0 project to increase cost-effectiveness and financial performance by implementing cutting-edge technologies. IoT sensors, data analytics, and automation technologies are all integrated into the project. Specific financial goals are specified throughout project planning, such as lowering operational costs, maximizing resource utilization, and raising profitability.

The project team carries out a thorough financial analysis, calculating the initial investment necessary for technology deployment, ongoing maintenance expenses, and prospective cost savings. They create a thorough project budget that accounts for costs related to infrastructure improvements, software licenses, training, and equipment purchases. Key performance indicators (KPIs) for financial objectives, such as return on investment (ROI), cost reductions, and revenue growth, are identified by the project team. The team tracks costs, revenue sources, and possible savings as it executes the project, comparing financial performance to the set KPIs. They analyze cost patterns, spot possible waste, and put cost-cutting plans into action using data analytics technologies. The organization meets its financial goals by skillfully managing project finances, realizing cost savings, boosting profitability, and maximizing the return on investment in the Industry 4.0 effort.

7.15 ENHANCEMENT THROUGH IR 4.0

An IR 4.0 project is not abandoned after it is put into practice, in contrast to conventional IT projects. If a failure happens in production for any reason, there should be a plan for how the IoT will be serviced and how the failover mechanics will be rectified. Until Internet industries evolve in the autonomous manufacturing business, IoT projects will have a long tail that extends into maintenance and failover of IoT even after it is cut over to production. Because IoT will have an impact on all automated systems, networks, databases, storage, and IT operations functions, the project timeline will be longer as a result (Mantravadi et al., 2020). Collaboration, mobility, and social networking not just within teams but also among stakeholders are some of the digital trends that are increasingly responding to a dynamic and changing world.

Tools for collaboration encourage teams to work together to complete a task. However, in today's world, project managers need to be cautious when choosing the best method for contacting stakeholders (Doroshuk, 2021). A variety of tools, including phone conversations, webinars, and document collaboration software, should be used by the project manager. Before selecting a stakeholder group to support a project, the project manager should always attempt to obtain formal approval from that organization, as shown in Figure 7.15.

The project's success is greatly influenced by the project manager's ability to select the best communication tools for communicating with teams and stakeholders. A few means for enhancement, shown in Figure 7.2, are discussed subsequently.

7.15.1 Cyber-Physical System (CPS)

The Cyber-Physical System (CPS) is another technological advancement that has contributed to the enhancement of the PMBOK knowledge areas. CPS combines physical components such as sensors, actuators, and controllers with computational components such as software and

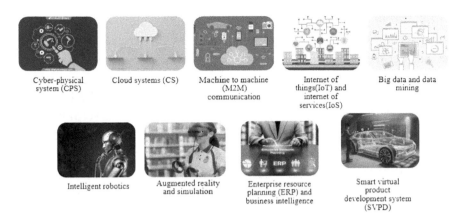

Cyber-physical system (CPS)	Cloud systems (CS)	Machine to machine (M2M) communication	Internet of things(IoT) and internet of services(IoS)	Big data and data mining

Intelligent robotics	Augmented reality and simulation	Enterprise resource planning (ERP) and business intelligence	Smart virtual product development system (SVPD)

FIGURE 7.15   Enhancement Through Industrial Revolution (IR 4.0).

communication networks. The integration of these components creates a system that can interact with the physical world and respond to changes in real time. The use of CPS in project management has led to the evolution of project quality management, project risk management, and project resource management. With the integration of CPS, project managers can now monitor and control project quality in real time. CPS sensors can detect changes in project performance, and this information can be transmitted to the project manager who can take corrective action immediately. PS also enhances project risk management by providing real-time information about potential risks (Lee, 2015). With CPS, project managers can detect potential risks early and take action to mitigate them before they escalate into major problems.

CPS can also facilitate effective resource management by monitoring resource utilization and providing real-time data on resource availability. Another area that CPS has enhanced in project management is project communication management. The use of CPS has facilitated the sharing of real-time data and information, which has improved communication between team members. With CPS, team members can collaborate more effectively and communicate seamlessly, leading to better project outcomes. Finally, CPS has also impacted project procurement management by enabling the automation of procurement processes. With CPS, project managers can now automate procurement processes such as ordering, invoicing, and payment processing, reducing the time and effort required for procurement activities (Palma et al., 2019).

7.15.2 Cloud Systems (CS)

Software as a Service (SaaS), Platform as a Service (PaaS), and Infrastructure as a Service (IaaS) are just a few examples of cloud computing services that can be delivered via a network and are examples of cloud systems or cloud computing technologies (i.e., the Internet) (Adjepon-Yamoah, 2019). Cloud systems can enhance several knowledge areas within the PMBOK framework, including the following:

7.15.2.1 Project Integration Management

- Cloud systems can provide a centralized platform for project management, where all project-related information, data, and communications can be accessed and managed from a single location.

- Cloud systems can improve project integration management by providing greater visibility into project status and progress.

7.15.2.2 Project Scope Management

- Cloud systems can facilitate scope management by providing tools for collaborative requirement gathering and analysis, and by enabling real-time tracking of scope changes and updates.

7.15.2.3 Project Time Management

- Cloud systems can support time management by providing tools for scheduling, resource allocation, and task tracking.

- With real-time data and analytics, project managers can better optimize their schedules and mitigate potential delays.

7.15.2.4 Project Cost Management

- Cloud systems can improve cost management by providing real-time data on project expenses and budgets, as well as tools for cost analysis and forecasting.

7.15.2.5 Project Quality Management

- Cloud systems can support quality management by providing tools for quality planning, assurance, and control.

- Cloud systems can also enable collaboration among team members and stakeholders in identifying and resolving quality issues.

7.15.2.6 Project Risk Management
- Cloud systems can enhance risk management by providing tools for risk identification, assessment, and mitigation.

- Cloud systems can also enable real-time monitoring of risks and proactive response planning.

7.15.2.7 Project Communication Management
- Cloud systems can improve communication management by providing a platform for real-time collaboration and communication among project team members and stakeholders.

7.15.2.8 Project Stakeholder Management
- Cloud systems can support stakeholder management by providing tools for stakeholder engagement and collaboration.

- Cloud systems can also enable real-time tracking of stakeholder expectations and feedback.

Overall, cloud systems can provide several benefits to project management, including improved efficiency, collaboration, and real-time decision-making. By leveraging cloud systems, project managers can optimize their project management processes and improve project outcomes (Sepasgozar et al., 2019).

7.15.3 Machine-to-Machine (M2M) Communication
The term "machine-to-machine," or M2M, is a general one that can be applied to any technology that allows networked devices to share information and carry out tasks without the need for human intervention (Frederico, 2021). Machine-to-machine (M2M) communication can enhance several knowledge areas within the PMBOK framework, including the following:

7.15.3.1 Project Integration Management
- M2M communication can provide a centralized platform for project management, where machines can communicate with each other and exchange information, data, and status updates in real time.

- This can improve project integration management by providing greater visibility into project status and progress.

7.15.3.2 Project Scope Management

- M2M communication can facilitate scope management by enabling machines to collaborate in requirement gathering, analysis, and monitoring.

- For instance, sensors can monitor the environment and feed data back to project managers, who can then adjust the project scope accordingly.

7.15.3.3 Project Time Management

- M2M communication can support time management by enabling machines to communicate and collaborate on scheduling, resource allocation, and task tracking.

- This can help project managers optimize their schedules and mitigate potential delays.

7.15.3.4 Project Cost Management

- M2M communication can improve cost management by enabling machines to communicate and share data on project expenses and budgets, as well as tools for cost analysis and forecasting.

7.15.3.5 Project Quality Management

- M2M communication can support quality management by enabling machines to collaborate on quality planning, assurance, and control.

- For example, machines can be programmed to monitor and detect quality issues and alert project managers for timely resolution.

7.15.3.6 Project Risk Management

- M2M communication can enhance risk management by enabling machines to communicate and share data on risk identification, assessment, and mitigation.

- For example, machines can be programmed to monitor environmental and other risks and alert project managers for proactive response planning.

7.15.3.7 Project Communication Management

- M2M communication can improve communication management by enabling machines to communicate with each other in real time, without the need for human intervention.

- This can improve the speed and accuracy of communication and collaboration among project team members and stakeholders.

7.15.3.8 Project Stakeholder Management

- M2M communication can support stakeholder management by enabling machines to communicate with stakeholders in real time.

- For example, machines can be programmed to send automated updates and alerts to stakeholders, keeping them informed and engaged throughout the project.

Overall, M2M communication can provide several benefits to project management, including improved efficiency, collaboration, and real-time decision-making. By leveraging M2M communication, project managers can optimize their project management processes and improve project outcomes (Janczewski and Ward, 2019).

7.15.4 Internet of Things (IoT) and Internet of Services (IoS)

Most new applications are still developed and tested independently by information technology (IT), and even when the application is delivered to the customer's production site, IT specialists must still provide services on their own. This is true even though agile development offers a more collaborative approach to projects (Prasher and Onu, 2020).

IoT projects cannot be completed using this standard operating procedure because IoT is so closely related to business activities that it is impossible to isolate the software and hardware from the actual operating environment. IT projects and industry operations must constantly collaborate on IoT throughout the project's life span. The Internet of Things (IoT) and Internet of Services (IoS) can enhance several knowledge areas within the PMBOK framework, including the following:

7.15.4.1 Project Integration Management

- IoT and IoS can provide a centralized platform for project management, where data from multiple sources can be integrated and analyzed in real time.

- This can improve project integration management by providing greater visibility into project status and progress.

7.15.4.2 Project Scope Management

- IoT and IoS can facilitate scope management by providing tools for collaborative requirement gathering and analysis.

- For example, sensors and other IoT devices can be used to gather data for project requirements, while IoS platforms can be used to analyze and process this data.

7.15.4.3 Project Time Management

- IoT and IoS can support time management by providing tools for scheduling, resource allocation, and task tracking.

- For example, IoT devices can be used to track progress on tasks, while IoS platforms can be used to optimize scheduling and resource allocation.

7.15.4.4 Project Cost Management

- IoT and IoS can improve cost management by providing real-time data on project expenses and budgets, as well as tools for cost analysis and forecasting.

- For example, IoT devices can be used to track expenses, while IoS platforms can be used to analyze this data and make cost projections.

7.15.4.5 Project Quality Management

- IoT and IoS can support quality management by providing tools for quality planning, assurance, and control.

- For example, IoT devices can be used to monitor quality metrics, while IoS platforms can be used to analyze this data and identify quality issues.

7.15.4.6 Project Risk Management

- IoT and IoS can enhance risk management by providing tools for risk identification, assessment, and mitigation.

- For example, IoT devices can be used to monitor environmental and other risks, while IoS platforms can be used to analyze this data and make risk projections.

7.15.4.7 Project Communication Management

- IoT and IoS can improve communication management by providing a platform for real-time collaboration and communication among project team members and stakeholders.

- For example, IoS platforms can be used to facilitate communication between IoT devices and project team members.

7.15.4.8 Project Stakeholder Management

- IoT and IoS can support stakeholder management by providing tools for stakeholder engagement and collaboration.

- For example, IoS platforms can be used to facilitate communication between IoT devices and stakeholders and to track stakeholder expectations and feedback.

Overall, IoT and IoS can provide several benefits to project management, including improved efficiency, collaboration, and real-time decision-making (Frederico, 2021). By leveraging these technologies, project managers can optimize their project management processes and improve project outcomes.

7.15.5 Big Data and Data Mining

Big data and data mining can enhance several knowledge areas within the PMBOK framework, including the following:

7.15.5.1 Project Integration Management

- Big data and data mining can provide insights into project status and progress, which can be used to integrate data from multiple sources in real time.

- This can improve project integration management by providing greater visibility into project status and progress.

7.15.5.2 Project Scope Management
- Big data and data mining can facilitate scope management by providing tools for requirement gathering and analysis.

- For example, data mining algorithms can be used to identify patterns in data that can help with requirement gathering and analysis.

7.15.5.3 Project Time Management
- Big data and data mining can support time management by providing tools for scheduling, resource allocation, and task tracking.

- For example, big data can be used to identify patterns and trends that can help with project scheduling and resource allocation.

7.15.5.4 Project Cost Management
- Big data and data mining can improve cost management by providing real-time data on project expenses and budgets, as well as tools for cost analysis and forecasting.

- For example, big data can be used to identify cost-saving opportunities, while data mining can be used to analyze this data and make cost projections.

7.15.5.5 Project Quality Management
- Big data and data mining can support quality management by providing tools for quality planning, assurance, and control.

- For example, big data can be used to identify patterns and trends in quality metrics, while data mining can be used to analyze this data and identify quality issues.

7.15.5.6 Project Risk Management
- Big data and data mining can enhance risk management by providing tools for risk identification, assessment, and mitigation.

- For example, big data can be used to monitor environmental and other risks, while data mining can be used to analyze this data and make risk projections.

7.15.5.7 Project Communication Management

- Big data and data mining can improve communication management by providing a platform for real-time collaboration and communication among project team members and stakeholders.

- For example, data mining algorithms can be used to identify patterns in stakeholder communication that can help with communication planning.

7.15.5.8 Project Stakeholder Management

- Big data and data mining can support stakeholder management by providing tools for stakeholder engagement and collaboration.

- For example, big data can be used to identify stakeholder expectations and feedback, while data mining can be used to analyze this data and make stakeholder projections.

Overall, big data and data mining can provide several benefits to project management, including improved efficiency, collaboration, and real-time decision-making (Errezgouny and Cherkaoui, 2022). By leveraging these technologies, project managers can optimize their project management processes and improve project outcomes (Nino et al., 2015; Ghrachorloo et al., 2023).

7.15.6 Intelligent Robotics

Robotic process automation (RPA) is a technology that was created to increase productivity and efficiency while reducing expenses, which is why project managers find it to be of immeasurable value. Automating processes frees up people's time and attention, enabling them to produce worthwhile output for the majority of their working hours (Paschek et al., 2022). Intelligent robotics can enhance several knowledge areas within the PMBOK framework, including the following:

7.15.6.1 Project Integration Management

- Intelligent robotics can provide a centralized platform for project management, where data from multiple sources can be integrated and analyzed in real time.

- This can improve project integration management by providing greater visibility into project status and progress.

7.15.6.2 Project Scope Management
- Intelligent robotics can facilitate scope management by providing tools for collaborative requirement gathering and analysis.

- For example, robots can be used to gather data on project requirements and analyze this data using machine learning algorithms.

7.15.6.3 Project Time Management
- Intelligent robotics can support time management by providing tools for scheduling, resource allocation, and task tracking.

- For example, robots can be used to automate tasks and optimize scheduling and resource allocation.

7.15.6.4 Project Cost Management
- Intelligent robotics can improve cost management by providing real-time data on project expenses and budgets, as well as tools for cost analysis and forecasting.

- For example, robots can be used to automate tasks and reduce labor costs, while machine learning algorithms can be used to analyze data and make cost projections.

7.15.6.5 Project Quality Management
- Intelligent robotics can support quality management by providing tools for quality planning, assurance, and control.

- For example, robots can be used to perform quality inspections and identify defects, while machine learning algorithms can be used to analyze this data and identify quality issues.

7.15.6.6 Project Risk Management
- Intelligent robotics can enhance risk management by providing tools for risk identification, assessment, and mitigation.

- For example, robots can be used to perform tasks in hazardous environments, while machine learning algorithms can be used to analyze data and make risk projections.

7.15.6.7 Project Communication Management
- Intelligent robotics can improve communication management by providing a platform for real-time collaboration and communication among project team members and stakeholders.
- For example, robots can be used to facilitate communication between team members and stakeholders.

7.15.6.8 Project Stakeholder Management
- Intelligent robotics can support stakeholder management by providing tools for stakeholder engagement and collaboration.
- For example, robots can be used to provide demonstrations and simulations to stakeholders, while machine learning algorithms can be used to analyze stakeholder feedback.

Overall, intelligent robotics can provide several benefits to project management, including improved efficiency, collaboration, and real-time decision-making (Frederico, 2021). By leveraging these technologies, project managers can optimize their project management processes and improve project outcomes. However, it is important to note that the use of intelligent robotics may require additional resources and expertise, and careful planning and implementation are necessary to ensure success (Jeremy, 2019).

7.15.7 Augmented Reality and Simulation
When display-based systems integrate actual and virtual visuals and are interactive in real time, this is known as augmented reality (AR) in simulation education (Paulauskas et al., 2023). Augmented reality and simulation can enhance several knowledge areas within the PMBOK framework, including the following:

7.15.7.1 Project Integration Management
- Augmented reality and simulation can provide a centralized platform for project management, where data from multiple sources can be integrated and analyzed in real time.
- This can improve project integration management by providing greater visibility into project status and progress.

7.15.7.2 Project Scope Management
- Augmented reality and simulation can facilitate scope management by providing tools for requirement gathering and analysis.

- For example, augmented reality can be used to visualize project requirements, while simulation can be used to test different scenarios and identify potential issues.

7.15.7.3 Project Time Management
- Augmented reality and simulation can support time management by providing tools for scheduling, resource allocation, and task tracking.

- For example, simulation can be used to test different scheduling scenarios and optimize resource allocation.

7.15.7.4 Project Cost Management
- Augmented reality and simulation can improve cost management by providing real-time data on project expenses and budgets, as well as tools for cost analysis and forecasting.

- For example, simulation can be used to analyze data and make cost projections.

7.15.7.5 Project Quality Management
- Augmented reality and simulation can support quality management by providing tools for quality planning, assurance, and control.

- For example, augmented reality can be used to visualize quality standards and identify quality issues, while simulation can be used to test quality controls and identify potential issues.

7.15.7.6 Project Risk Management
- Augmented reality and simulation can enhance risk management by providing tools for risk identification, assessment, and mitigation.

- For example, simulation can be used to test different scenarios and identify potential risks, while augmented reality can be used to visualize and assess risk factors.

7.15.7.7 Project Communication Management

- Augmented reality and simulation can improve communication management by providing a platform for real-time collaboration and communication among project team members and stakeholders.

- For example, augmented reality can be used to provide visual demonstrations and simulations to stakeholders.

7.15.7.8 Project Stakeholder Management

- Augmented reality and simulation can support stakeholder management by providing tools for stakeholder engagement and collaboration.

- For example, augmented reality can be used to visualize project progress and provide updates to stakeholders, while simulation can be used to identify stakeholder expectations and feedback.

Overall, augmented reality and simulation can provide several benefits to project management, including improved efficiency, collaboration, and real-time decision-making. By leveraging these technologies, project managers can optimize their project management processes and improve project outcomes (Calderon-Hernandez and Brioso, 2018). However, it is important to note that the use of augmented reality and simulation may require additional resources and expertise, and careful planning and implementation are necessary to ensure success.

7.15.8 Enterprise Resource Planning (ERP) and Business Intelligence

ERP software programs enable businesses to unify all business operations into a single system. Planning, inventory purchases, sales, marketing, finances, human resources, customer relationship management, and other activities will all be integrated.

ERP is a key integration tool in business that increases daily work in a variety of ways and simplifies operations. ERP coordinates all transactions and workflow, making it simple to get the data that is crucial for business decisions. ERP systems offer insights into customer communications and interactions, ensuring that the customer is effectively always catered to. ERP systems will save costs through quicker development times and more effective distribution and supply chain management (Aldossari and Mokhtar, 2020). Enterprise resource planning (ERP) and business intelligence (BI) can enhance several knowledge areas within the PMBOK framework, including the following:

7.15.8.1 Project Integration Management

- ERP systems can provide a centralized platform for project management, where data from multiple sources can be integrated and analyzed in real time.

- This can improve project integration management by providing greater visibility into project status and progress.

7.15.8.2 Project Scope Management

- ERP systems can help project managers identify and define project scope more effectively by providing tools for requirement gathering and analysis.

- For example, an ERP system can help to identify the resources required for a particular project, as well as the timelines and milestones needed to achieve project objectives.

- BI tools can help to analyze project scope data, identify potential issues, and suggest solutions.

7.15.8.3 Project Time Management

- ERP systems can support time management by providing tools for scheduling, resource allocation, and task tracking.

- For example, an ERP system can help to schedule tasks and allocate resources to meet project deadlines.

- BI tools can help to monitor project timelines and identify potential delays, allowing project managers to take corrective action where necessary.

7.15.8.4 Project Cost Management

- ERP systems can improve cost management by providing real-time data on project expenses and budgets, as well as tools for cost analysis and forecasting.

- For example, an ERP system can help track project expenses and forecast future costs based on historical data.

- BI tools can help to identify cost-saving opportunities and suggest alternative strategies to reduce project costs.

7.15.8.5 Project Quality Management

- ERP systems can support quality management by providing tools for quality planning, assurance, and control.

- For example, an ERP system can help to track project quality metrics and identify areas where quality standards are not being met.

- BI tools can help to analyze quality data and identify potential issues, allowing project managers to take corrective action where necessary.

7.15.8.6 Project Risk Management

- ERP systems can enhance risk management by providing tools for risk identification, assessment, and mitigation.

- For example, an ERP system can help to identify potential risks associated with a project, as well as the impact of these risks on project outcomes.

- BI tools can help to monitor project risk data and identify potential issues, allowing project managers to take corrective action where necessary.

7.15.8.7 Project Communication Management

- ERP systems can improve communication management by providing a platform for real-time collaboration and communication among project team members and stakeholders.

- For example, an ERP system can help to share project data and updates with stakeholders in real time.

- BI tools can help to monitor communication data and identify potential issues, allowing project managers to take corrective action where necessary.

7.15.8.8 Project Stakeholder Management

- ERP systems can support stakeholder management by providing tools for stakeholder engagement and collaboration.

- For example, an ERP system can help to identify key stakeholders and their interests in a project, as well as their level of involvement.

- BI tools can help to monitor stakeholder data and identify potential issues, allowing project managers to take corrective action where necessary.

Overall, ERP systems and BI tools can provide several benefits to project management, including improved efficiency, collaboration, and real-time decision-making. By leveraging these technologies, project managers can optimize their project management processes and improve project outcomes. However, it is important to note that the use of ERP systems and BI tools may require additional resources and expertise, and careful planning and implementation are necessary to ensure success (Khalilzadeh and Alikhani, 2020).

7.15.9 Smart Virtual Product Development System (SVPD)

Smart devices boost usefulness, enhance safety, foster interconnectedness, and develop new modes of communication. Smart products, however, inevitably make the concept, design, and production processes more difficult. The Industrial Revolution 4.0 (IR 4.0) is rapidly transforming industries and businesses around the world. It is a new era of digital transformation, characterized by the fusion of technologies such as artificial intelligence, robotics, the Internet of Things (IoT), and big data (Ahmed et al., 2020). As industries and businesses undergo this transformation, project management practices are also evolving to accommodate the changing landscape.

A Smart Virtual Product Development System (SVPD) can enhance several knowledge areas within the PMBOK framework, including the following:

7.15.9.1 Project Scope Management

- SVPD can help project managers define and validate project scope more effectively by providing tools for virtual prototyping and simulation.

- For example, SVPD can help project teams design and test new product concepts virtually, reducing the need for physical prototypes and shortening the development cycle. This can also improve the accuracy of project scope estimation and reduce the risk of scope creep.

7.15.9.2 Project Time Management

- SVPD can support time management by providing tools for virtual testing and optimization.

- For example, SVPD can help project teams identify and resolve potential design issues before physical prototyping, reducing the need for rework and delays. This can also help to accelerate the development cycle and reduce time-to-market.

7.15.9.3 Project Cost Management

- SVPD can improve cost management by providing tools for virtual prototyping and optimization.

- For example, SVPD can help project teams identify potential design flaws and cost-saving opportunities before physical prototyping, reducing the need for expensive rework and material waste. This can also help to optimize project budgets and reduce project costs.

7.15.9.4 Project Quality Management

- SVPD can support quality management by providing tools for virtual testing and validation.

- For example, SVPD can help project teams simulate and test product performance under different conditions, ensuring that quality standards are met before physical prototyping. This can also help to reduce the risk of product recalls and improve customer satisfaction.

7.15.9.5 Project Risk Management

- SVPD can enhance risk management by providing tools for virtual testing and optimization.

- For example, SVPD can help project teams identify and mitigate potential design risks before physical prototyping, reducing the risk of product failures and liability issues. This can also help to optimize project outcomes and reduce the risk of project failure.

7.15.9.6 Project Communication Management

- SVPD can improve communication management by providing a platform for virtual collaboration and feedback.

- For example, SVPD can help project teams share and review virtual prototypes and design concepts, improving communication and collaboration among team members and stakeholders. This can also help to reduce miscommunication and ensure that project requirements are met.

7.15.9.7 Project Stakeholder Management

- SVPD can support stakeholder management by providing a platform for virtual collaboration and feedback.

- For example, SVPD can help project teams engage and collaborate with stakeholders during the design and development process, ensuring that their needs and requirements are met. This can also help to improve stakeholder satisfaction and reduce the risk of project delays and failures.

Overall, SVPD can provide several benefits to project management, including improved efficiency, collaboration, and quality. By leveraging these technologies, project managers can optimize their product development processes and improve project outcomes. However, it is important to note that the use of SVPD may require additional resources and expertise, and careful planning and implementation are necessary to ensure success (Ahmed et al., 2019).

7.16 DISCUSSION

IR 4.0 is changing the way project managers integrate and manage project components. With the rise of big data and the IoT, project managers can gather and analyze vast amounts of data in real time, allowing them to make more informed decisions. In addition, the integration of new technologies such as artificial intelligence and machine learning is enabling project managers to automate tasks and optimize project workflows, leading to more efficient project delivery. IR 4.0 is enabling project managers to better manage project scope by providing new tools and technologies to collect and analyze data (Smith, 2016). For example, machine learning algorithms can help project managers analyze large data sets to identify patterns and trends, which can be used to refine project scope. In addition, technologies such as augmented and virtual reality are enabling project managers to visualize project components and identify potential scope issues before they occur (El Yamami et al., 2017).

IR 4.0 is enhancing project time management by providing new tools and technologies to track project schedules and milestones. For example, project managers can use project management software that automatically updates schedules in real time, allowing project managers to track progress and adjust schedules as necessary. In addition, new technologies such as drones and robotics can help project managers monitor progress on construction projects, providing real-time data to help them manage project schedules (Nauman and Piracha, 2016). IR 4.0 enables project managers to better manage project costs by providing new tools and technologies to analyze and optimize project budgets. For example, machine learning algorithms can analyze project data to identify cost-saving opportunities, while robotics and automation can help reduce labor costs.

In addition, blockchain technology can be used to track project costs and ensure transparency in project finances. IR 4.0 is enhancing project quality management by providing new tools and technologies to monitor and improve project quality (Frederico, 2021). For example, machine learning algorithms can analyze project data to identify potential quality issues, while sensors and the IoT can provide real-time data to help project managers identify quality issues as they arise. In addition, technologies such as 3D printing and additive manufacturing are enabling project managers to create high-quality project components more efficiently and cost-effectively.

IR 4.0 is enhancing project resource management by providing new tools and technologies to manage project resources more efficiently. For example, project management software can help project managers allocate resources more effectively, while robotics and automation can help optimize resource utilization. In addition, technologies such as the IoT can provide real-time data on resource utilization, enabling project managers to make data-driven decisions about resource allocation. IR 4.0 is enhancing project communication management by providing new tools and technologies to facilitate communication and collaboration among project stakeholders. For example, project management software can provide a centralized platform for project communication, while virtual and augmented reality technologies can enable remote collaboration among project teams (Mikhieieva and Waidmann, 2017). In addition, new communication technologies such as chatbots and voice assistants can help project managers automate routine communication tasks, freeing up time for more strategic communication efforts. Project cost management involves planning, estimating, budgeting, financing, funding, managing, and controlling costs so that the project can be completed within the approved budget. In IR 4.0, project cost management has been enhanced using advanced cost management tools that incorporate data analytics and artificial intelligence. These tools enable project managers to make more accurate cost estimates and forecasts, track and analyze project costs in real time, and identify cost-saving opportunities. Project quality management involves planning, executing, and controlling the quality of the project deliverables to meet the stakeholders' expectations (Carden et al., 2021).

In IR 4.0, project quality management has been enhanced using advanced quality management tools that incorporate machine learning, big data analytics, and sensors. These tools enable project managers to collect real-time data on project quality, identify quality issues early, and take corrective action before the project is completed. Project resource management involves planning, acquiring, and managing the resources required for the project's successful completion. In IR 4.0, project resource management has been enhanced using advanced resource management tools that incorporate data analytics, artificial intelligence, and machine learning. These tools enable project managers to optimize resource utilization, identify resource constraints early, and allocate resources more effectively (Pandi-Perumal et al., 2015).

Project communication management involves planning, executing, and monitoring the project's communication activities to ensure that the project stakeholders receive the right information at the right time. In IR 4.0, project communication management has been enhanced using advanced communication tools that incorporate social media, instant messaging, and collaboration platforms. These tools enable project managers to communicate with stakeholders more effectively, share project information in real time, and collaborate with team members more efficiently. Project risk management involves identifying, analyzing, and responding to project risks to minimize their impact on the project (Alwaly and Alawi, 2020). In IR 4.0, project risk management has been enhanced using advanced risk management tools that incorporate predictive analytics, artificial intelligence, and machine learning. These tools enable project managers to identify potential risks early, predict their likelihood and impact, and develop effective risk response strategies (Lee, 2015).

Project procurement management involves planning, managing, and controlling the procurement of goods and services required for the project's successful completion. In IR 4.0, project procurement management has been enhanced using advanced procurement tools that incorporate data analytics, artificial intelligence, and machine learning. These tools enable project managers to optimize procurement processes, identify cost-saving opportunities, and manage supplier relationships more effectively (Alwaly and Alawi, 2020). Project stakeholder management involves identifying, analyzing, and managing stakeholders' needs and expectations to ensure their support for the project. In IR 4.0, project stakeholder management has been enhanced by the use of advanced stakeholder management tools that incorporate social media, instant messaging, and collaboration platforms (Cardona-Meza and Olivar-Tost, 2017). These tools enable project managers to engage with stakeholders more effectively, identify stakeholder concerns early, and develop effective stakeholder engagement strategies.

The integration of the 13 knowledge areas outlined in PMBOK (Project Management Body of Knowledge) forms the foundation of a comprehensive project management plan. When coupled with the implementation of circular economy principles and the integration of strategies from Industry 4.0, a synergistic effect emerges, resulting in a pronounced and direct impact. The seamless incorporation of circular economy practices and Industry 4.0 strategies amplifies the overall project's significance, transcending each PMBOK knowledge area. By infusing circular economy ideals and Industry 4.0 tactics into every knowledge area, the objective is to not only heighten project magnitude but also to attain sustainability within the current climate context, as shown in Figure 7.16. This combined approach directly bolsters sustainability outcomes, yielding a holistic result that aligns with the project's overarching goals and the imperatives of the prevailing climatic circumstances.

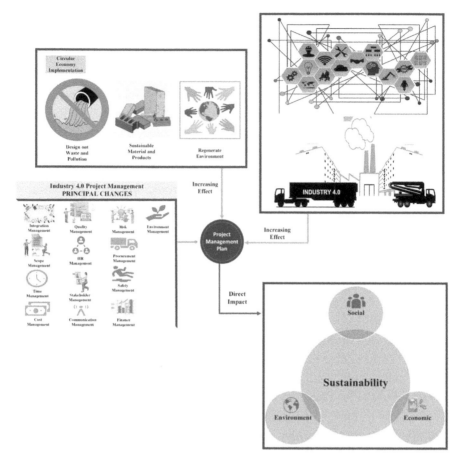

FIGURE 7.16   Project Management Framework in the Era of IR 4.0.

7.17 SUMMARY

IR 4.0 has revolutionized project management by introducing advanced technology, which has led to the evolution and enhancement of PMBOK knowledge areas. The nine PMBOK knowledge areas discussed earlier, namely project integration management, project scope management, project cost management, project quality management, project resource management, project communication management, project risk management, project procurement management, and project stakeholder management have all been enhanced by IR 4.0. The use of advanced technology such as data analytics, artificial intelligence, machine learning, and social media platforms has enabled project managers to improve their planning, monitoring, and controlling processes. These tools provide real-time data, making it easier for project managers to make informed decisions, anticipate issues, and identify opportunities for improvement.

IR 4.0 has also improved project management using advanced project management software and tools, which allow project managers to manage multiple projects simultaneously, track progress, and communicate with team members in real time. These tools have also facilitated effective resource allocation, enhanced procurement processes, and improved stakeholder management. In summary, IR 4.0 has enhanced the PMBOK knowledge areas by introducing advanced technology that provides real-time data, facilitates effective resource allocation, and improves procurement processes. Project managers can now make informed decisions, anticipate issues, and identify opportunities for improvement, which ultimately leads to better project outcomes. Therefore, project managers must continue to embrace and leverage the benefits of IR 4.0 to optimize project management and achieve project success. IR 4.0 is not just a trend but it is a new way of life, and those who are adaptable and innovative will thrive in this new era of project management.

REFERENCES

Abdilahi, S. M., Fakunle, F. F. & Fashina, A. 2020. Exploring the extent to which project scope management processes influence the implementation of telecommunication projects. *PM World Journal*, IX(5), 1–17.

Adjepon-Yamoah, D. E. 2019. Reactive middleware for effective requirement change management of cloud-based global software development. *Software Engineering for Resilient Systems: 11th International Workshop, SERENE 2019, Naples, Italy, September 17, Proceedings 11, 2019*. Springer, 46–66.

Ahmed, M. B., Majeed, F., Sanin, C. & Szczerbicki, E. 2020. Enhancing product manufacturing through smart virtual product development (SVPD) for Industry 4.0. *Cybernetics and Systems*, 51, 246–257.

Ahmed, M. B., Sanin, C. & Szczerbicki, E. 2019. Smart virtual product development (SVPD) to enhance product manufacturing in Industry 4.0. *Procedia Computer Science*, 159, 2232–2239.

Alaloul, W. S., Liew, M. S., Zawawi, N. A. W. A. & Mohammed, B. S. 2018. Industry revolution IR 4.0: Future opportunities and challenges in construction industry. *MATEC Web of Conferences*. EDP Sciences, 02010.

Aldossari, S. & Mokhtar, U. A. 2020. A model to adopt enterprise resource planning (ERP) and business intelligence (BI) among Saudi SMEs. *International Journal of Innovation*, 8, 305–347.

Al-Rubaiei, Q. H. S., Nifa, F. A. A. & Musa, S. 2018. Project scope management through multiple perspectives: A critical review of concepts. *AIP Conference Proceedings*. AIP Publishing LLC, 020025.

Alwaly, K. A. & Alawi, N. A. 2020. Factors affecting the application of project management knowledge guide (PMBOK® GUIDE) in construction projects in Yemen. *International Journal of Construction Engineering and Management*, 9, 81–91.

Ashaari, M. A., Singh, K. S. D., Abbasi, G. A., Amran, A. & Liebana-Cabanillas, F. J. 2021. Big data analytics capability for improved performance of higher education institutions in the era of IR 4.0: A multi-analytical SEM & ANN perspective. *Technological Forecasting and Social Change*, 173, 121119.

Baker, B. 2018. Project quality management practice & theory. *American Journal of Management*, 18, 10–17.

Blocher, E., Stout, D., Juras, P. & Smith, S. 2019. *Cost Management (A Strategic Emphasis) 8e*. McGraw-Hill Education.

Calderon-Hernandez, C. & Brioso, X. 2018. Lean, BIM and augmented reality applied in the design and construction phase: A literature review. *International Journal of Innovation, Management and Technology*, 9, 60–63.

Carden, L., Kovach, J. V. & Flores, M. 2021. Enhancing human resource management in process improvement projects. *Organizational Dynamics*, 50, 100776.

Cardona-Meza, L. S. & Olivar-Tost, G. 2017. Modeling and simulation of project management through the PMBOK® standard using complex networks. *Complexity*, 2017(4), 1–12.

Chin, L. S. & Hamid, A. R. A. 2015. The practice of time management on construction project. *Procedia Engineering*, 125, 32–39.

Chung, K. S. K. & Crawford, L. 2016. The role of social networks theory and methodology for project stakeholder management. *Procedia-Social and Behavioral Sciences*, 226, 372–380.

Demirkesen, S. & Ozorhon, B. 2017. Impact of integration management on construction project management performance. *International Journal of Project Management*, 35, 1639–1654.

Doroshuk, H. 2021. Prospects and efficiency measurement of artificial intelligence in the management of enterprises in the energy sector in the era of Industry 4.0. *Polityka Energetyczna*, 24.

El Yamami, A., Ahriz, S., Mansouri, K., Qbadou, M. & Illoussamen, E. 2017. Representing IT projects risk management best practices as a metamodel. *Engineering, Technology & Applied Science Research*, 7, 2062–2067.

Errezgouny, A. & Cherkaoui, A. 2022. Contribution in big data projects management. *E3S Web of Conferences*. EDP Sciences, 01066.

Frederico, G. F. 2021. Project management for supply chains 4.0: A conceptual framework proposal based on PMBOK methodology. *Operations Management Research*, 14, 434–450.

Ghrachorloo, N., Nouri, F., Javanmardi, M. & Taghizadeh, H. 2023. Big data mining in the analysis of factors affecting the occurrence of natural gas incidents in East Azerbaijan province (IRAN). *Journal of Applied Research on Industrial Engineering,* 10(4), 599–614.

Gładysz, B., Skorupka, D., Kuchta, D. & Duchaczek, A. 2015. Project risk time management–a proposed model and a case study in the construction industry. *Procedia Computer Science,* 64, 24–31.

Gutierrez Aguilar, F. 2022. *Procurement Processes in Construction in Europe: An Assessment Following Project Procurement Management PMBOK Approach.* Universitat Politècnica de Catalunya.

Hansen, D. R., Mowen, M. M. & Heitger, D. L. 2021. *Cost Management.* Cengage Learning.

Hardin, B. & McCool, D. 2015. *BIM and Construction Management: Proven Tools, Methods, and Workflows.* John Wiley & Sons.

Hermano, V. & Martín-Cruz, N. 2019. Expanding the knowledge on project management standards: A look into the PMBOK® with dynamic lenses. In *Project Management and Engineering Research: AEIPRO 2017.* Springer, 19–34.

Hessing, M. & Summerville, T. 2014. *Canadian Natural Resource and Environmental Policy: Political Economy and Public Policy.* UBC Press.

Jamali, G. & Oveisi, M. 2016. A study on project management based on PMBOK and PRINCE2. *Modern Applied Science,* 10, 142–146.

Janczewski, L. J. & Ward, G. 2019. *IOT: Challenges in Information Security Training.* Kennesaw State University. https://digitalcommons. kennesaw.edu/ccerp/2019/education/3/

Jeremy, T. Z. K. 2019. *Fourth Industrial Revolution in the Singapore Construction Industry: Challenges, Strategies and Impact on the Project Management Body of Knowledge.* https://scholarbank.nus.edu. sg/handle/10635/220272

Kamaruzaman, M., Hamid, R., Mutalib, A. & Rasul, M. 2019. Comparison of engineering skills with IR 4.0 skills. *International Journal of Online and Biomedical Engineering (iJOE),* 15(10), 15–28.

Kerzner, H. 2017. *Project Management: A Systems Approach to Planning, Scheduling, and Controlling.* John Wiley & Sons.

Khalilzadeh, M. & Alikhani, A. M. 2020. The effects of PMBOK knowledge areas on the phases of ERP implementation. *Industrial Engineering & Management Systems,* 19, 242–253.

Langston, C. 2013. Development of generic key performance indicators for PMBOK using a 3D project integration model. *The Australasian Journal of Construction Economics and Building,* 13, 78–91.

Lee, E. A. 2015. The past, present and future of cyber-physical systems: A focus on models. *Sensors,* 15, 4837–4869.

Mantravadi, S., Schnyder, R., Møller, C. & Brunoe, T. D. 2020. Securing IT/OT links for low power IIoT devices: Design considerations for Industry 4.0. *IEEE Access,* 8, 200305–200321.

Maskuriy, R., Selamat, A., Maresova, P., Krejcar, O. & David, O. O. 2019. Industry 4.0 for the construction industry: Review of management perspective. *Economies*, 7, 68.

Mikhieieva, O. & Waidmann, M. 2017. Communication management tools for managing projects in an intercultural environment. *PM World Journal*, 6, 1–15.

Moustafaev, J. 2014. *Project Scope Management: A Practical Guide to Requirements for Engineering, Product, Construction, IT and Enterprise Projects*. CRC Press.

Nauman, S. & Piracha, M. 2016. Project stakeholder management: A developing country perspective. *Journal of Quality and Technology Management*, 12, 1–24.

Nino, M., Blanco, J. M. & Illarramendi, A. 2015. Business understanding, challenges and issues of big data analytics for the servitization of a capital equipment manufacturer. *2015 IEEE International Conference on Big Data (Big Data)*. IEEE, 1368–1377.

Palma, F. E., Fantinato, M., Rafferty, L. & Hung, P. C. 2019. Managing scope, stakeholders and human resources in cyber-physical system development. *ICEIS*, 2, 36–47.

Pandi-Perumal, S. R., Akhter, S., Zizi, F., Jean-Louis, G., Ramasubramanian, C., Edward Freeman, R. & Narasimhan, M. 2015. Project stakeholder management in the clinical research environment: How to do it right. *Frontiers in Psychiatry*, 6, 71.

Park, S. & Huh, J.-H. 2018. Effect of cooperation on manufacturing it project development and test bed for successful Industry 4.0 project: Safety management for security. *Processes*, 6, 88.

Paschek, D., Luminosu, C.-T. & Ocakci, E. 2022. Industry 5.0 challenges and perspectives for manufacturing systems in the society 5.0. *Sustainability and Innovation in Manufacturing Enterprises: Indicators, Models and Assessment for Industry 5.0*, 17–63.

Paulauskas, L., Paulauskas, A., Blažauskas, T., Damaševičius, R. & Maskeliūnas, R. 2023. Reconstruction of industrial and historical heritage for cultural enrichment using virtual and augmented reality. *Technologies*, 11, 36.

Pheng, L. S. & Pheng, L. S. 2018. Project procurement management. *Project Management for the Built Environment: Study Notes*, 177–193.

Prasher, V. S. & Onu, S. 2020. The Internet of Things (IoT) upheaval: Overcoming management challenges. *The Journal of Modern Project Management*, 8.

Qureshi, A. H., Alaloul, W. S., Manzoor, B., Musarat, M. A., Saad, S. & Ammad, S. 2020. Implications of machine learning integrated technologies for construction progress detection under Industry 4.0 (IR 4.0). *2020 Second International Sustainability and Resilience Conference: Technology and Innovation in Building Designs (51154)*. IEEE, 1–6.

Rustamova, S. 2021. Application of PMI PMBOK standard in the development of the construction cost management projects plan. *International Journal of Trends in Business Administration*, 11.

Samáková, J., Sujanová, J. & Koltnerová, K. 2013. Project communication management in industrial enterprises. *7th European Conference on Information Management and Evaluation, ECIME.* The University of Gdańsk, Faculty of Management, 155–163.

Sepasgozar, S. M., Karimi, R., Shirowzhan, S., Mojtahedi, M., Ebrahimzadeh, S. & McCarthy, D. 2019. Delay causes and emerging digital tools: A novel model of delay analysis, including integrated project delivery and PMBOK. *Buildings,* 9, 191.

Shahroom, A. A. & Hussin, N. 2018. Industrial revolution 4.0 and education. *International Journal of Academic Research in Business and Social Sciences,* 8, 314–319.

Sherwani, F., Asad, M. M. & Ibrahim, B. S. K. K. 2020. Collaborative robots and industrial revolution 4.0 (ir 4.0). *2020 International Conference on Emerging Trends in Smart Technologies (ICETST).* IEEE, 1–5.

Smith, P. 2014. Project cost management–Global issues and challenges. *Procedia-Social and Behavioral Sciences,* 119, 485–494.

Smith, P. 2016. Project cost management with 5D BIM. *Procedia-Social and Behavioral Sciences,* 226, 193–200.

Taghipour, M., Hoseinpour, Z., Mahboobi, M., Shabrang, M. & Lashkarian, T. 2015. Construction projects risk management by risk allocation approach using PMBOK standard. *Journal of Applied Environmental and Biological Sciences,* 5, 323–329.

Teo, T., Unwin, S., Scherer, R. & Gardiner, V. 2021. Initial teacher training for twenty-first century skills in the Fourth Industrial Revolution (IR 4.0): A scoping review. *Computers & Education,* 170, 104223.

Tsaramirsis, G., Kantaros, A., Al-Darraji, I., Piromalis, D., Apostolopoulos, C., Pavlopoulou, A., Alrammal, M., Ismail, Z., Buhari, S. M. & Stojmenovic, M. 2022. A modern approach towards an Industry 4.0 model: From driving technologies to management. *Journal of Sensors,* 2022.

Willumsen, P., Oehmen, J., Stingl, V. & Geraldi, J. 2019. Value creation through project risk management. *International Journal of Project Management,* 37, 731–749.

Zachko, O., Golovatyi, R. & Kobylkin, D. 2019. Models of safety management in development projects. *2019 IEEE 14th International Conference on Computer Sciences and Information Technologies (CSIT).* IEEE, 81–84.

Zaouga, W., Rabai, L. B. A. & Alalyani, W. R. 2019. Towards an ontology based-approach for human resource management. *Procedia Computer Science,* 151, 417–424.

Index

0–9

3 dimensional (3D), 14, 20, 26, 38, 41,
42, 51, 52–54, 57, 62, 100, 101,
103, 205
3D construction printing (3DCP), 53
3D printer (3DP), 53

A

artificial intelligence (AI), 5, 8, 52–53,
54, 68, 88, 102, 104, 105, 108, 113,
143–145, 151, 153, 156, 171, 202,
204, 206–208
Assets Information Requirement
(AIR), 43
augmented and virtual reality (AVR),
7, 8, 20, 51, 205
augmented reality (AR), 26, 51, 52,
156, 166, 197–199, 206

B

BIM Execution Plan (BEP), 43
building information modeling (BIM),
8, 9, 12, 20, 36–38, 40–48, 51, 56,
57, 58, 60, 62, 64–65, 144, 145
Building Research Establishment
Environmental Assessment Method
(BREEAM), 148
business intelligence (BI), 86, 110,
199, 200

C

carbon dioxide (CO_2), 123, 127
computer-aided design (CAD), 40, 60

cost-benefit analysis (CBA), 160
Cyber-Physical Production System
(CPPS), 11, 21, 56
cyber-physical systems (CPS), 26,
56–57, 65, 79, 186, 187

D

Digital Supply Network (DSN), 20, 21

E

earned value analysis (EVA or EVM),
73, 74, 160
Employer's Information Requirement
(EIR), 43
enterprise resource planning (ERP),
199–203
environmental impact assessment
(EIA), 94, 134, 135

F

facility managers (FMs or FM), 32

G

Global Positioning System (GPS), 39
gross domestic product (GDP), 26, 39

I

Identity of Things (IDOT), 21
indoor environmental quality (IEQ),
128, 140
Industrial Internet of Things (IIOT), 21
Industrial Revolution 1.0 (IR.0), 4, 9, 10

Industrial Revolution 2.0 (IR 2.0), 4, 11
Industrial Revolution 3.0 (IR 3.0), 4, 11
Industrial Revolution 4.0 (IR 4.0), 4–6,
    8, 9–14, 19, 20, 21, 24, 26, 29, 31,
    37, 39, 59, 63, 67, 68, 73, 75–77,
    79–81, 88, 99–108, 110–114, 119,
    143, 145, 146, 150–152, 153, 154,
    156–158, 161–164, 166–169, 171,
    172–174, 176, 178, 179, 181, 182,
    184–187, 203, 204–209
Industrial Revolution (IR), 4, 5, 8, 9, 10,
    18, 26, 37, 59, 60, 61, 65, 99, 101,
    102, 105, 108, 109, 143, 150, 152
Industry 4.0 (I 4.0), 8, 9, 11–13, 15–27,
    29–32, 39, 59, 61–62, 64–68,
    75–77, 79, 82, 112, 113, 143–145,
    153–159, 161, 163–165, 167, 171,
    173, 176–180, 184–186, 207
information technology (IT), 63, 191
Integrated Facilities Management
    (IFM), 33
integrated project management (IPM or
    IPD), 43–45
integrated workplace management
    platform or systems (IWMS), 19,
    32, 33
Internet of Services (IoS), 20, 191
Internet of Things (IoT), 8, 18, 20, 55,
    99, 102, 108, 113, 143, 144, 191, 203

K

key performance indicators (KPIs), 86,
    95, 96, 98, 185

L

life cycle assessment (LCA), 97, 98

M

machine learning (ML), 88, 144
machine-to-machine (M2M), 20, 189–191

P

Platform as a Service (PaaS), 188
Product Development to Services, 21

program evaluation and review
    technique (PERT), 74
Project Management Body of
    Knowledge (PMBOK), 68, 80,
    89, 109–114, 150, 171, 186, 188,
    189, 191, 193, 195, 197, 200, 202,
    207–209
Project Management Institute (PMI),
    71, 109
project risk management (PRM),
    171–174, 187, 189, 190, 193, 194,
    196, 198, 201, 203, 206–208

Q

quality function deployment (QFD), 162

R

research and development (R&D), 23,
    75, 86
return on investment (ROI), 185, 186
robotic process automation (RPA), 153,
    159, 161, 173, 196

S

sick building syndrome (SBS), 127
small and midsize enterprises (SME), 38
Smart Virtual Product Development
    System (SVPD), 202, 203, 204
Software as a Service (SaaS), 188
Sustainable Development Goals
    (SDGs), 29, 143

U

universal basic income (UBI), 101, 107
unmanned aerial vehicles (UAVs), 39

V

virtual reality (VR), 8, 19, 20, 51, 156,
    166, 205

W

work breakdown structure (WBS), 74,
    156